Maths Frameworking

3rd edition

Kevin Evans, Keith Gordon,
Trevor Senior, Brian Speed,
Chris Pearce

Contents

How to use this book

Learning objectives

See what you are going to cover and what you should already know at the start of each chapter.

About this chapter

Find out the history of the maths you are going to learn and how it is used in real-life contexts.

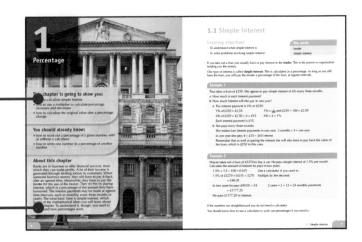

Key words

The main terms used are listed at the start of each topic and highlighted in the text the first time they come up, helping you to master the terminology you need to express yourself fluently about maths. Definitions are provided in the glossary at the back of the book.

Worked examples

Understand the topic before you start the exercises, by reading the examples in blue boxes. These take you through how to answer a question step by step.

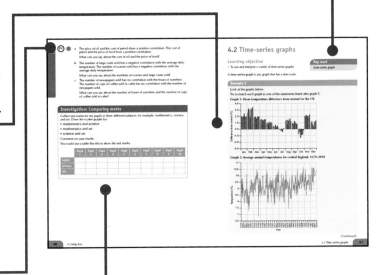

Skills focus

Practise your problem-solving, mathematical reasoning and financial skills.

Take it further

Stretch your thinking by working through the **Investigation**, **Problem solving**, **Challenge** and **Activity** sections. By tackling these you are working at a higher level.

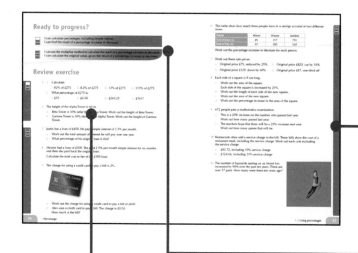

Progress indicators

Track your progress with indicators that show the difficulty level of each question.

Ready to progress?

Check whether you have achieved the expected level of progress in each chapter. The statements show you what you need to know and how you can improve.

Review questions

The review questions bring together what you've learnt in this and earlier chapters, helping you to develop your mathematical fluency.

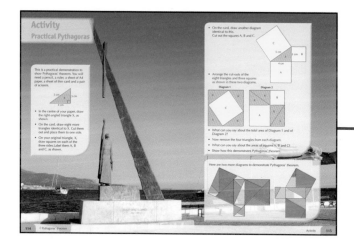

Activity pages

Put maths into context with these colourful pages showing real-world situations involving maths. You are practising your problem-solving, reasoning and financial skills.

Interactive book, digital resources and videos

A digital version of this Pupil Book is available, with interactive classroom and homework activities, assessments, worked examples and tools that have been specially developed to help you improve your maths skills. Also included are engaging video clips that explain essential concepts, and exciting real-life videos and images that bring to life the awe and wonder of maths.

Find out more at www.collins.co.uk/connect

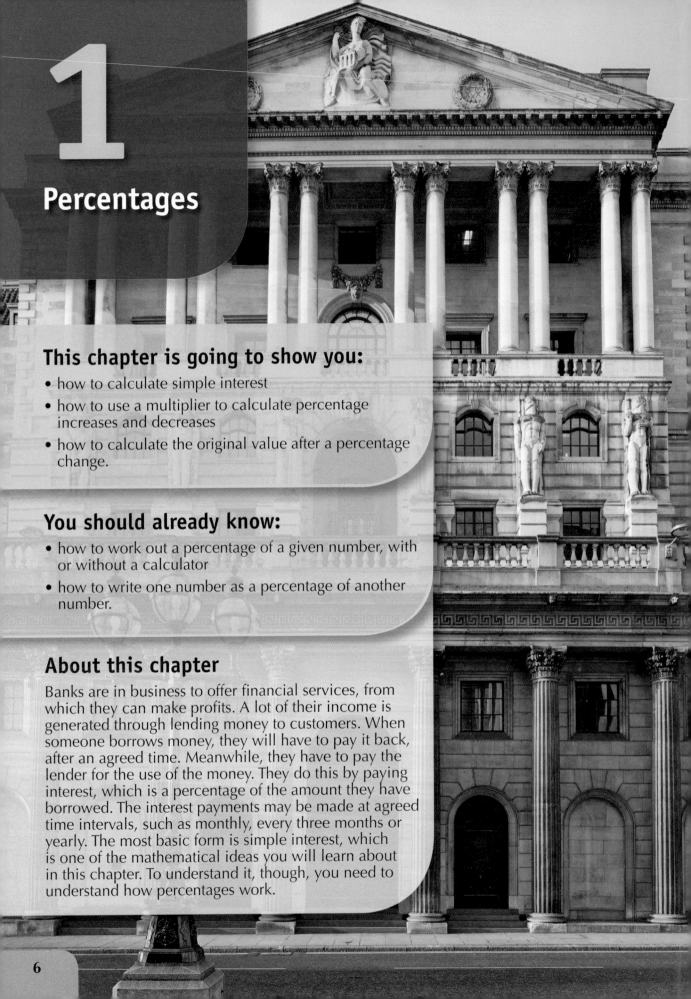

1

Percentages

This chapter is going to show you:

- how to calculate simple interest
- how to use a multiplier to calculate percentage increases and decreases
- how to calculate the original value after a percentage change.

You should already know:

- how to work out a percentage of a given number, with or without a calculator
- how to write one number as a percentage of another number.

About this chapter

Banks are in business to offer financial services, from which they can make profits. A lot of their income is generated through lending money to customers. When someone borrows money, they will have to pay it back, after an agreed time. Meanwhile, they have to pay the lender for the use of the money. They do this by paying interest, which is a percentage of the amount they have borrowed. The interest payments may be made at agreed time intervals, such as monthly, every three months or yearly. The most basic form is simple interest, which is one of the mathematical ideas you will learn about in this chapter. To understand it, though, you need to understand how percentages work.

1.1 Simple interest

Learning objectives

- To understand what simple interest is
- To solve problems involving simple interest

Key words

lender

simple interest

If you take out a loan you usually have to pay interest to the **lender**. This is the person or organisation lending you the money.

One type of interest is called **simple interest**. This is calculated as a percentage. As long as you still have the loan, you will pay the lender a percentage of the loan, at regular intervals.

Example 1

Tina takes a loan of £250. She agrees to pay simple interest of 6% every three months.

a How much is each interest payment?

b How much interest will she pay in one year?

a The interest payment is 6% of £250.

    1% of £250 = £2.50 $1\% = \frac{1}{100}$ and £250 ÷ 100 = £2.50

    6% of £250 = £2.50 × 6 = £15   6% = 6 × 1%

    Each interest payment is £15.

b She pays every three months.

    She makes four interest payments in one year.   3 months × 4 = one year

    In one year she pays 4 × £15 = £60 interest.

    Remember that as well as paying the interest she will also have to pay back the value of the loan, which is £250 in this case.

Example 2

Wayne takes out a loan of £3270 to buy a car. He pays simple interest of 1.5% per month. Calculate the amount of interest he pays in two years.

    1.5% = 1.5 ÷ 100 = 0.015

    1.5% of £3270 = 0.015 × 3270   Multiply by the decimal. Use a calculator.

                = £49.05

    In two years he pays £49.05 × 24.   2 years = 2 × 12 = 24 monthly payments

               = £1177.20

    He pays £1177.20 in interest.

If the numbers are straightforward you do not need a calculator.

You should know how to use a calculator to work out percentages if you need to.

Example 3

Lucy has a loan of £750. She pays interest of £25.50 per month. What is the rate of simple interest?

$$\frac{25.50}{750} = 0.034$$ Divide the interest by the amount of the loan.

$0.034 \times 100 = 3.4\%$ Multiply the decimal by 100 to change it to a percentage.

Exercise 1A

1 Work out these percentages. Do not use a calculator.

    a 25% of £300   **b** 10% of £45   **c** 40% of £1000   **d** 75% of £600

    e 30% of £2000   **f** 13% of £300   **g** 82% of £200   **h** 150% of £80

2 Use a calculator to work out these percentages.

    a 13% of £85   **b** 7% of £425   **c** 23% of £6500   **d** 3.5% of £230

    e 1.4% of £620   **f** 38% of £560   **g** 3.8% of £1320   **h** 0.5% of £325

FS 3 Sam takes a loan of £400.

He pays simple interest of 4% per month for nine months.

    a Work out the amount of interest Sam pays each month.

    b Work out the total amount of interest Sam pays.

FS 4 Harry pays 2.5% monthly interest on a loan of £3800.

He pays interest for one year.

    a Work out the amount of interest he pays each month.

    b Work out the total interest he has paid after one year.

FS 5 Jenny takes a loan of £5400.

She pays 3.2% simple interest monthly, for two years, and then she pays back the loan.

    a Calculate how much interest she pays each month.

    b Calculate how much interest she pays altogether.

FS **6** Aaron has a loan of £730.

He pays simple interest of 0.7% per month for eight months.

Calculate the total amount of interest Aaron pays.

FS **7** Shaun pays 8% per year on a loan of £3200.

He pays interest for seven years.

Work out how much interest he pays all together.

FS **8** Sally pays 0.5% per week interest on a loan of £60.

How much interest has she paid after 20 weeks?

PS **9** Cameron takes a loan of £5400 and pays 1.4% simple interest every month for six months.

Mary takes a loan of £3600 and pays 2.8% interest every month for nine months.

Who pays more interest, Cameron or Mary? Justify your answer.

FS **10** Graham has a loan of £600.

He pays £10.80 simple interest per month.

Work out the rate of interest.

FS **11** Amy pays £56 monthly interest on a loan of £4000.

Work out the rate of interest.

FS **12** Myra pays £11.52 simple interest per month on a loan of £640.

Work out the rate of interest.

PS **13** Jack pays 1.6% simple interest monthly on a loan of £800 for one year.

a Work out his total interest payment over the year.

b Work out his total interest payment as a percentage of the original loan.

Challenge: Using a formula

This is a formula you can use to work out simple interest payments.

$$I = \frac{PRT}{100}$$

where P is the initial loan, R is the percentage rate of interest, T is the number of payments and I is the total interest paid.

A If you take out a loan of £500 at 2% per month for six months, then $P = 500$, $R = 2$ and $T = 6$.

Use the formula to work out the total interest paid in this case.

B Work out the total simple interest paid on:

a a loan of £250 at 2.5% monthly interest for 12 months

b a loan of £1500 at 1.5% monthly interest for 10 months

c a loan of £3000 at 9% yearly interest for five years.

1.2 Percentage increases and decreases

Learning objectives

- To calculate the result of a percentage increase or decrease
- To choose the most appropriate method to calculate a percentage change

Key words

decrease	increase
multiplier	

A percentage change may be:

- an **increase** if the new value is larger than the original value
- a **decrease** if the new value is smaller that the original value.

There are several methods that you can use to calculate the result of a percentage change.

The **multiplier** method is often the most efficient. You just multiply the original value by an appropriate number, to calculate the result of the percentage change.

Example 4

Alicia buys a bicycle for £350. A year later the value of the bicycle has fallen by 20%.

Calculate the new value.

The original value was 100%.

You need to find 100% − 20% = 80% of the original price.

The value is 0.8 × £350 = £280

80% = $\frac{4}{5}$ or 0.8. This is the multiplier.

You could also work out $\frac{4}{5}$ × 350 to get the answer.

You might find the fraction method easier, if you do not have a calculator.

Example 5

Shaun puts £438 in a bank account. After one year it has earned 3% interest.

Work out the total amount in the account.

The interest must be added to the original.

You need to find 100% + 3% = 103%.

103% = 103 ÷ 100 = 1.03

Multiply by 1.03.

The new total is £438 × 1.03 = £451.14.

For this problem, it is sensible to use a calculator.

If you use the multiplier method, the multiplier will be:

- larger than 1 for an increase
- smaller than 1 for a decrease.

The box below will remind you.

To increase by 15%:	the multiplier is 100% + 15% = 115% = 1.15
To decrease by 15%:	the multiplier is 100% − 15% = 85% = 0.85

Example 6

The price of a car is decreased from £8490 to £7750.

Calculate the percentage decrease.

$\dfrac{\text{new price}}{\text{original price}} = \dfrac{7750}{8490}$ Write the new price as a fraction of the original price.

$= 0.9128\ldots$ This is $7750 \div 8490$.

$= 91.3\%$ The new price is 91.3% of the original price.

The reduction is $100\% - 91.3\% = 8.7\%$.

Example 7

An electricity bill increases from £285.79 to £299.42.

Calculate the percentage increase.

$\dfrac{\text{new bill}}{\text{original bill}} = \dfrac{299.42}{285.79}$ Write the new bill as a fraction of the original bill.

$= 1.0476\ldots$ This is $299.42 \div 285.79$

$= 104.8\%$ The new bill is 104.8% of the original bill.

The increase is $104.8\% - 100\% = 4.8\%$.

Exercise 1B

1   Peter has £400 in a savings account.

How much will he have if his savings:

a   increase   b   decrease   c   increase   d   decrease
    by 25%   by 25%   by 14%   by 14%?

2   Decrease each price by 15%.

a   Jacket, £160   b   Shoes, £90
c   Handbag, £75   d   Jeans, £48

3 Increase each bill by 9%.

    a Electricity, £325.40 **b** Gas, £216.53 **c** Rent, £475.50 **d** Council tax, £781.20

4 The price of a cooker was £585.

Work out the new price after:

£585

    a an increase of 3%   **b** a decrease of 13%

    c an increase of 23% **d** a decrease of 2.3%.

5 **a** When Sam was calculating a percentage change she multiplied the original value by 1.07.

    State whether this was an increase or a decrease and what the percentage change was.

    b When Phil was calculating a percentage change he multiplied the original value by 0.7.

    State whether this was an increase or a decrease and what the percentage change was.

6 The value of an antique painting is £5200.

Work out the new value if it increases by:

    a 20%   **b** 90%   **c** 110%   **d** 135%.

7 The rent for a flat increases from £320 to £336.

Work out the percentage increase.

8 Work out the percentage change in each price.

    a An increase from £280 to £305   **b** An increase from £975 to £1125

    c An increase from £76.50 to £105.50   **d** An increase from £37 to £87

9 Work out the percentage change in each price.

    a A decrease from £40 to £30   **b** A decrease from £90 to £80

    c A decrease from £480 to £465   **d** A decrease from £960 to £290

10 The price of a pair of shoes was reduced from £89 to £54.

Work out the percentage decrease.

11 The price of a house was £80 000.

Ten years later the price is £196 000.

Calculate the percentage increase in the price.

12 The mass of a child increases from 32.5 kg to 41.6 kg.

Work out the percentage increase.

 13 Peter has a rectangular piece of card with sides of length 16 cm and 20 cm.

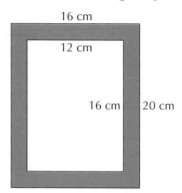

16 cm

12 cm

16 cm   20 cm

He cuts off a 2 cm border all around the edge.

The length and the width are both reduced by 20%.

Is Peter correct? Justify your answer.

Challenge: Population change

The population of a town increased by 20% from 1980 to 1990.

A In 1980 the population was 30 000. Work out the population in 1990.

The population of the town increased by 10% from 1990 to 2000.

B Work out the population in 2000.

The population of the town decreased by 5% from 2000 to 2010.

C Work out the population in 2010.

D Work out the percentage increase from 1980 to 2010.

1.3 Calculating the original value

Learning objective

- Given the result of a percentage change, to calculate the original value

Key word

original value

The number of swans in a wildlife reserve has increased by 20% since last year. There are now 450 swans. How many were there last year?

It would be incorrect to reduce 450 by 20% and get $450 \times 0.8 = 360$.

This is because the 20% relates to the **original value**, which is the number last year, not the number now.

The correct answer is that last year there were 375 swans.

This is because if you increase 375 by 20% you get $375 \times 1.2 = 450$.

In this section you will learn how to answer questions like this.

Example 8

Last year, the number of daily visitors to Hastings Museum and Art Gallery increased by 30% to 1066.

What was the number of visitors before the increase?

The multiplier for the increase is 1.3.	$100\% + 30\% = 130\% = 1.3$
Original number $\times 1.3 = 1066$	The result of the increase is 1066.
Original number $= \dfrac{1066}{1.3}$	Find the original number, by dividing by 1.3.
$= 820$	$1066 \div 1.3 = 820$

Example 9

In a survey, the number of butterflies seen in a wood was 150. This was a 40% reduction on the previous year. How many butterflies were there last year?

The multiplier for a 40% reduction is 0.6.	$100\% - 40\% = 60\% = 0.6$
Last year's number $\times 0.6 = 150$	
Last year's number $= \dfrac{150}{0.6} = 250$	Divide 150 by 0.6.

Prices often include a tax called value-added tax (VAT). This is a percentage of the basic cost.

You can use the multiplier method to find the cost before VAT is added.

Example 10

The price of a car repair, including 20% VAT, is £283.68.

Work out the price excluding VAT.

The multiplier for a 20% increase is 1.2. $100\% + 20\% = 120\% = 1.2$

Price excluding VAT $\times 1.2 = £283.68$

Price excluding VAT $= \dfrac{£283.68}{1.2} = £236.40$

Exercise 1C

1   Write down the multiplier for each percentage increase.

    a 10%   **b** 45%   **c** 87%   **d** 135%

2   Write down the multiplier for each percentage decrease.

    a 5%   **b** 42%   **c** 57%   **d** 85%

3   **a**   After a 10% increase, a price is now £275. Work out the original price.

    b   After a 20% increase, a price is now £540. Work out the original price.

    c   After a 30% increase, a price is now £234. Work out the original price.

    d   After a 5% increase, a price is now £756. Work out the original price.

4   **a**   After a 10% decrease, a price is now £630. Work out the original price.

    b   After a 25% decrease, a price is now £48. Work out the original price.

    c   After a 5% decrease, a price is now £621.25. Work out the original price.

    d   After an 80% decrease, a price is now £12. Work out the original price.

5   The number of trees in a wood is 575.
This is a 15% increase from five years ago.

Work out the number of trees that
were in the wood five years ago.

6   The amount of cereal in a packet has increased by 30%. The contents are now 780 g.
Work out the contents before the increase.

7   The rent on a house increases by 12% and is now £728.
Calculate the rent before the increase.

8   The price of concert tickets offered on a website is reduced by 10%. The price
now is £58.50.

Work out the original price.

9 Work out the original price of the coat.

COAT
30% reduction
NOW ONLY
£47.25

10 **a** A gym increased its membership fee by 25% to £160.
Work out the membership fee before the increase.

b The number of members in the gym decreased by 25% to 180.
Work out the membership numbers before the decrease.

FS **11** These are some prices, including VAT at 20%.
Work out the price excluding VAT in each case.

a Restaurant bill, £67.68   **b** Garage bill, £548.16   **c** Computer, £1054.80

FS **12** The rate of VAT on energy bills is 8%. These are some energy bills, including VAT.
Work out the bills before VAT was added.

a Electricity, £58.64   **b** Gas, £155.30   **c** Electricity, £304.05

FS **13** The standard rate of VAT in Ireland is 23%. These are the prices, including VAT, of some goods in Ireland.
Work out the prices excluding VAT.

a Camera, €350.55   **b** Printer, €102.64   **c** Furniture, €2287.80

14 The population of the UK in 2001 was 59 million. This was an increase of 55% from 1901.
Work out the population in 1901.

15 The number of members in a society has reduced by 65% over the last ten years.
Now there are 1498 members.
How many members were there in the same society ten years ago?

Financial skills: VAT

The rate of VAT changes from time to time and it is different in different countries.

The cost of an item, including VAT at 20%, is £564.

A What would it cost if the rate of VAT was 25%?

B What would it cost if the rate of VAT was 15%?

C In another country the equivalent price, including VAT, would be £554.60.
What is the rate of VAT in that country?

1.4 Using percentages

Learning objective

- To choose the correct calculation to work out a percentage

You now know how to make calculations involving percentages in a variety of situations. This section will give you practice in choosing the right calculation to use.

Example 11

The ratio of men to women in a keep-fit class is 4 : 3.

a Work out the percentage of the keep-fit class that is made up of men.

b What is the number of women, as a percentage of the number of men?

You can represent the situation with a diagram.

a Four out of every seven of the people are men. $4 + 3 = 7$

$\frac{4}{7}$ are men.

$\frac{4}{7} = \frac{4}{7} \times 100\% = 57\%$

$4 \div 7 = 0.5714\ldots$ Men Women

b There are three women to every four men.

There are $\frac{3}{4}$ as many women as there are men.

$\frac{3}{4} = 75\%$

Example 12

In a music examination, if you pass you may also get a distinction.

One year 256 pupils passed and 62 of them got a distinction.

109 pupils failed.

What percentage got a distinction:

a out of those who passed

b out of those who took the examination?

a 62 got a distinction, and 256 passed.

62 out of 256 $= \frac{62}{256} \times 100\%$

$= 24.2\%$ $62 \div 256 = 0.2421\ldots$

b 365 pupils took the examination.   That is $256 + 109$.

62 out of 365 $= \frac{62}{365} \times 100\% = 17.0\%$ $62 \div 365 = 0.1698\ldots$

1. a   One week last year, 93 people took a driving test and 37 passed.
        Work out the percentage that passed.
    b   The following week the number who passed decreased by 16%.
        How many passed in that week?

2. There are cars and vans in a car park. The ratio of cars to vans is 7 : 1.
    a   Work out the percentage of the vehicles that are cars.
    b   What is the number of cars as a percentage of the number of vans?

3. In an audience there are 80 men, 90 women, 60 girls and 20 boys.
    a   What percentage of the children are girls?
    b, What percentage of the adults are women?
    c   What percentage of the whole audience is female?

4. The number of visitors to a website each day increased from 4600 to 6400.
    Work out the percentage increase.

5. In 2000 there were 64 elephants in a wildlife
    reserve. In 2010 the number had increased by 25%.
    a   Work out the number of elephants in 2010.
    b   What percentage of the 2010 number is the
        2000 number?
    c   What percentage of the 2000 number is
        the 2010 number?

6. Mike and Joe play in a football team. In three
    seasons, Mike scored 40 goals and Joe scored 32 goals.
    a   Work out Joe's goals as a percentage of the total number of goals the two scored.
    b   Work out the number of goals Joe scored as a percentage of the number Mike scored.
    c   Work out Mike's total goals as a percentage of Joe's total goals.

7. These are the numbers of votes cast for four candidates in an election.

Candidate	Alan	Kirsty	Matt	Wendy
Votes	50	65	75	40

    a   What percentage of the votes did Kirsty get?
    b   What percentage of the votes that went to females did Wendy get?
    c   Write the number of votes that Alan got as a percentage of the number of votes
        that Kirsty got.

8. Some college students are either running or swimming
    for charity.
    The ratio of runners to swimmers is 3 : 2.
    a   What percentage of the students are runners?
    b   There are 144 runners. How many students are
        there altogether?
    c   What is the number of swimmers, as a percentage
        of the number of runners?

9 1340 people visited an amusement park on Friday.

    a On Saturday there were 30% more visitors than on Friday. How many visited on Saturday?

    b Of the total for Friday and Saturday, what percentage visited on Saturday?

    c On Friday there were 10% more than there were on Thursday. How many visited on Thursday?

(FS) 10 a The price of a laptop was £185. In a sale the price was reduced to £169. What was the percentage decrease?

    b The price of a television was originally £479. In the sale there was a 15% reduction. What was the sale price?

    c After a 30% reduction, the price of a radio was £58.80. Work out the original price.

(FS) 11 Here are some prices. The rate of VAT is 15%.

Work out the missing numbers.

	Item	Price before VAT (£)	Price including VAT (£)
	Fit gas cooker	85.00	97.75
a	Fix broken window	44.00	
b	Install computer		34.27
c	Service boiler	69.40	
d	Replace radiator		164.22

12 There are 60 children and teenagers living in a new housing estate. This is 20% of the total population.

    a How many people live in the estate altogether?

    b Work out the ratio of children and teenagers to adults.

Challenge: Different representations

Look at this information.

An animal refuge looks after dogs and cats. There are 48 dogs and 32 cats.

You could show that information in a number of ways.

- The ratio of dogs to cats is 3 : 2.
- 60% of the animals are dogs.
- The number of cats is $\frac{2}{3}$ of the number of dogs.

A Show that each of the statements above is correct.

B Here is some more information.

    A farmer has cows and sheep on his farm. He has 75 cows and 45 sheep.

    Write statements like the ones above, to show this information.

    Use ratios, fractions and percentages.

C Here is some more information.

    On an aeroplane there are 16 children and 224 adults.

    Write statements like the ones above, to show this information.

    Use ratios, fractions and percentages.

Ready to progress?

 I can calculate percentages, including simple interest.
I can find the result of a percentage increase or decrease.

 I can use the multiplier method to calculate the result of a percentage increase or decrease.
I can calculate the original value, given the result of a percentage increase or decrease.

Review questions

(FS) **1 a** Calculate:

     i 82% of £275   ii 8.2% of £275   iii 13% of £275   iv 113% of £275.

    b What percentage of £275 is:

     i £35   ii £6.50   iii £263.25   iv £361?

2 The height of the Alpha Tower is 60 m.

    a Beta Tower is 30% taller than Alpha Tower. Work out the height of Beta Tower.

    b Gamma Tower is 30% shorter than Alpha Tower. Work out the height of Gamma Tower.

3 Justin has a loan of £450. He pays simple interest of 1.5% per month.

    a Work out the total amount of interest he will pay over one year.

    b What percentage of his original loan is this?

(FS) **4** Maxine had a loan of £300. She paid 2.5% per month simple interest for six months and then she paid back the original loan.

    Calculate the total cost to her of the £300 loan.

(FS) **5** The charge for using a credit card to pay a bill is 2%.

    a Work out the charge for using a credit card to pay a bill of £600.

    b Alex uses a credit card to pay a bill. The charge is £0.50.
     How much is the bill?

6 This table show how much three people have in a savings account at two different times.

Name	Alison	Wayne	Jasmine
End of June (£)	85	157	755
End of July (£)	97	300	320

Work out the percentage increase or decrease for each person.

FS 7 Work out these sale prices.

   a Original price £75, reduced by 25%
   b Original price £820, cut by 10%
   c Original price £120, down by 60%
   d Original price £87, one-third off

8 Each side of a square is 8 cm long.

   a Work out the area of the square.
   Each side of the square is increased by 25%.
   b Work out the length of each side of the new square.
   c Work out the area of the new square.
   d Work out the percentage increase in the area of the square.

9 672 people pass a mathematics examination.

   a This is a 20% increase on the number who passed last year.
      Work out how many passed last year.
   b The teachers hope that there will be a 20% increase next year.
      Work out how many passes that will be.

FS 10 Restaurants often add a service charge to the bill. These bills show the cost of a restaurant meal, including the service charge. Work out each cost excluding the service charge.

   a £93.72, including 10% service charge
   b £124.66, including 15% service charge

PS 11 The number of buzzards nesting on an island has increased by 90% over the past ten years. There are now 57 pairs. How many were there ten years ago?

Challenge
Exponential growth

1 Steve has saved £1000. In future he plans to increase his savings by £1000 each year.
Copy and complete this table to show how much he will have.

Time	Start	1 year	2 years	3 years	4 years	5 years
Steve's amount (£)	1000	2000				

2 Sally has saved £1000. She wants to increase this by 50% each year.

a Show that the multiplier for a 50% increase is 1.5.

b Show that after one year Sally will have £1500.

c Multiply £1500 by 1.5 to work out how much Sally will have after two years.

d Work out how much Sally will have after three years.

e Copy and complete this table. Round the amounts to the nearest pound.

Time	Start	1 year	2 years	3 years	4 years	5 years
Sally's amount (£)	1000	1500				

f Copy and extend this graph. The years must go up to 5. The amount must go up to £8000.

g Plot Steve's values and Sally's values.
Join Steve's points with a straight line.
Join Sally's points with a smooth curve.

h Describe how Steve's line will continue if you add more years.

i Describe how Sally's line will continue if you add more years.

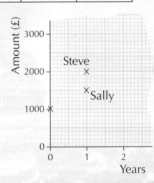

Sally's line is an example of exponential growth. The amount increases by a fixed percentage each year.

Here is another example.

3 Jordan has saved £1000 and plans to increase it by 25% each year.

  a Write down the multiplier for a 25% increase.

  b Copy and complete this table to show how much Jordan will have each year.

Time	Start	1 year	2 years	3 years	4 years	5 years
Jordan's amount (£)	1000					

  c Plot Jordan's values on your graph. Join them with a smooth curve.

  d How is Jordan's curve similar to Sally's curve?

  e How is Jordan's curve different from Sally's curve?

2

Equations and formulae

This chapter is going to show you:

- how to expand brackets and factorise algebraic expressions
- how to solve more complex equations
- how to rearrange formulae.

You should already know:

- how to collect like terms in an expression
- how to use one or two operations to solve equations
- how to substitute values into a formula
- what a highest common factor (HCF) is.

About this chapter

During World War 2, it was vital for all sides to keep up with communications, and to try to find out what each other were doing. One very important factor, for Britain and its allies, was the work done at Bletchley Park, where mathematicians worked to break the codes used to send messages. To do this, they had to be able to use algebraic rules to solve formulae. Algebraic rules are still used by computer programmers today.

The picture shows *Colossus*, the machine developed through the work of Alan Turing, the famous code-breaker. He is acknowledged to be a pioneer of theoretical computer science and artificial intelligence. *Colossus* is now considered to be the first digital computer.

In this chapter, you will learn about some of the basic ideas that were used by Alan Turing and his colleagues.

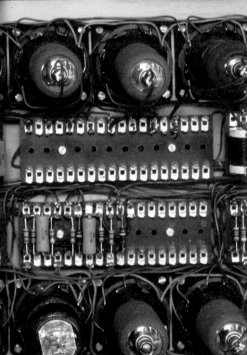

2.1 Multiplying out brackets

Learning objective

- To multiply out brackets

Key word

expand

You have seen before how to multiply out a term that includes brackets. If there is a number outside the brackets you multiply each term inside the brackets by that number.

This is also called **expanding** brackets.

For example, $5(8 + 3)$ is the same as $5 \times 8 + 5 \times 3$.

$$5(8 + 3) = 5 \times 11 = 55$$

$$5 \times 8 + 5 \times 3 = 40 + 15 = 55$$

Both have the same answer.

This also works if you replace some of the numbers inside or outside the brackets by letters.

Example 1

Multiply out the brackets in these expressions.

a $x(x - 5)$ **b** $3(6y + 5)$ **c** $-2(t - 3)$

a $x(x - 5) = x^2 - 5x$ $x \times x = x^2$ and $x \times -5 = -5x$

b $3(6y + 5) = 18y + 15$ $3 \times 6y = 18y$ and $3 \times 5 = 15$

c $-2(t - 3) = -2t + 6$ -2 is outside the brackets. $-2 \times t = -2t$ and $-2 \times -3 = 6$

Notice that in part **c**, because the product of -2 and -3 is 6, the second term is $+ 6$.

When you have more than one set of brackets you can remove them and then simplify the remaining expression by collecting like terms.

Example 2

Simplify these expressions by expanding the brackets and collecting like terms.

a $2(t - 3) + 3(t - 4)$ **b** $d(d + 1) - d(d - 2)$

a $2(t - 3) + 3(t - 4) = 2t - 6 + 3t - 12$ Multiply out each set of brackets separately.

$\qquad\qquad\qquad\qquad = 5t - 18$ $2t + 3t = 5t$ and $-6 - 12 = -18$

b $d(d + 1) - d(d - 2) = d^2 + d - d^2 + 2d$ In the second set of brackets, $-d \times d = -d^2$ and $-d \times -2 = 2d$.

$\qquad\qquad\qquad\qquad\qquad = 3d$ The d^2 terms cancel out.

Exercise 2A

1 Write each expression more concisely.

  a $2 \times 3a$ **b** $5 \times 4b$ **c** $6 \times 6c$ **d** $10d \times 5$

2 Multiply out the brackets.

  a $2(a + 5)$ **b** $4(t + 7)$ **c** $3(c - 6)$ **d** $6(a - 1.5)$

  e $t(4 + t)$ **f** $a(3 - a)$ **g** $x(5 - x)$ **h** $y(y + 4)$

3 Multiply out the brackets.

  a $4(2n + 1)$ **b** $3(2t + 7)$ **c** $3(5k - 2)$ **d** $10(2x - 5)$

  e $2(4k + 3)$ **f** $5(2n + 7)$ **g** $5(3 + 2d)$ **h** $3(3 - 4x)$

  i $1.5(2c + 9)$ **j** $2(2z + 3)$ **k** $5(6 - 3y)$ **l** $0.2(15 - 5x)$

4 Multiply out the brackets.

  a $-3(a + 2)$ **b** $-3(m + 5)$ **c** $-3(5k - 2)$ **d** $-10(2x - 5)$

  e $k(5 + k)$ **f** $f(8 - f)$ **g** $-x(4 - x)$ **h** $t(3.5 + t)$

5 Expand the brackets and simplify these expressions by collecting like terms.

  a $4(t + 2) + 3$ **b** $8(g + 1) - 5$ **c** $5(t - 2) - 3$ **d** $7(m - 3) + 13$

  e $2(x + 4) + x$ **f** $4(y + 1) - 2y$ **g** $6(n - 2) - 5n$ **h** $3(k - 2) + 2k$

  i $5 + 2(x + 1)$ **j** $9 + 3(a - 2)$ **k** $x + 3(x + 2)$ **l** $2t + 3(t - 1)$

6 Expand the brackets and collect like terms.

  a $2(d + 3) - (d + 1)$ **b** $3(x + 4) - (x + 4)$ **c** $3(y - 2) - (y - 1)$

7 Simplify these expressions as much as possible.

  a $5t - 2t$ **b** $4t - 2(t + 1)$ **c** $5t - 2(t - 1)$ **d** $5t - 2(t - 4)$

8 Simplify each expression as much as possible.

  a $2(x + 1) + 2(x + 3)$ **b** $x(x - 1) + x(x + 3)$ **c** $2(x - 1) + 2(x - 3)$

  d $3(t + 1) + 2(t - 3)$ **e** $w(w + 1) + w(w - 1)$ **f** $3(y - 1) + 4(y + 1)$

Challenge: Consecutive numbers

The numbers n, $n + 1$ and $n + 2$ are three consecutive integers.

A What are the integers, if $n = 7$?

B An expression for 'the first number plus twice the second number' is $n + 2(n + 1)$.

  a Simplify this expression as much as possible.

  b Check that your answer to **a** gives the correct answer, if $n = 7$.

C An expression for 'the first number plus twice the second number plus three times the third number' is $n + 2(n + 1) + 3(n + 2)$.

  a Simplify this expression as much as possible.

  b Check that your answer to **a** gives the correct answer when $n = 7$.

2.2 Factorising algebraic expressions

Learning objective

• To factorise expressions

To **factorise** an expression, you do the opposite of multiplying out the brackets. You need to look for a factor that is common to all the terms and take it outside the brackets.

• If you multiply out $4(x - 5)$ you get $4x - 20$.
• If you factorise $4x - 20$ you get $4(x - 5)$.

In this case, 4 is a common factor of 4 and 20 so it can be put outside the brackets.

Example 3

Factorise each expression.　**a** $12x + 18$　**b** $9a - 12b - 15$

a The HCF of 12 and 18 is 6.

$12x + 18 = 6(2x + 3)$

b A factor of 9, 12 and 15 is 3.

$9a - 12b - 15 = 3(3a - 4b - 5)$

2, 3 and 6 are factors. Always choose the highest.

6 goes outside the brackets.

The numbers inside are $12 \div 6 = 2$ and $18 \div 6 = 3$.

3 is their only common factor, apart from 1.

Divide 9, 12 and 15 by 3.

Look again at part **a**.

You could take out 2 as a factor and write $12x + 18 = 2(6x + 9)$.

This is not factorised completely because 6 and 9 have a common factor of 3. To factorise the expression completely you must take out 3 as a factor, as well.

$2(6x + 9) = 2 \times 3(2x + 3) = 6(2x + 3)$

This is the same answer. Always check that you have factorised as much as possible.

Remember that multiplying out brackets and factorising are inverse operations.

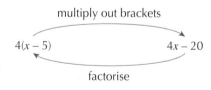

$4(x - 5)$ — multiply out brackets → $4x - 20$ — factorise

Exercise 2B

1　Work out the highest common factor of the numbers in each pair.

　a　15 and 10　　**b**　4 and 20　　**c**　18 and 24　　**d**　12 and 20

2　Factorise each expression.

　a　$2x + 6$　　　**b**　$3d - 12$　　**c**　$4y + 8$　　**d**　$5e - 20$

　e　$2x - 24$　　**f**　$6t + 18$　　**g**　$6n - 48$　　**h**　$8 + 4c$

　i　$21 - 7t$　　**j**　$8 - 8x$　　　**k**　$3 + 3t$　　　**l**　$5m + 60$

3 Factorise each expression as much as possible.

    a $4x + 6$ **b** $9x - 6$ **c** $12y + 16$ **d** $10y - 15$

    e $12z - 9$ **f** $20t + 30$ **g** $20t - 12$ **h** $12 + 16c$

    i $18 - 12x$ **j** $30 - 18t$ **k** $25 + 20t$ **l** $45m + 27$

4 Factorise each expression as much as possible.

    a $2a + 4b$ **b** $6a + 2b$ **c** $5c + 20d$ **d** $18c - 12d$

    e $24x + 30y$ **f** $6x - 8y$ **g** $40x + 24y$ **h** $14x - 7y$

5 Simplify each expression by collecting like terms.

Then factorise it.

    a $5a + 5 + 3a + 11$ **b** $7x - 6 + x - 2$ **c** $12y - 3 + 18y - 7$

    d $12x - 3 - 8x - 9$ **e** $x + 12 + 9x - 7$ **f** $8 - 3y + 10 - 3y$

(PS) **6** For each shape, write an expression for the perimeter.

Then simplify it as much as possible. Factorise it if you can.

a
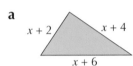

b
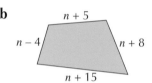

c
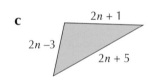

d
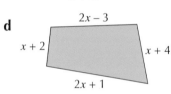

(MR) **7** Four integers are n, $n + 6$, $n + 12$ and $n + 18$.

    a Write down the four numbers, when:

        i $n = 10$ **ii** $n = 30$ **iii** $n = 101$.

    b Write down an expression for the sum of the four numbers.

        Simplify your expression and factorise it as much as possible.

(MR) **8** Three numbers are x, $2x - 1$ and $3x + 5$.

    a Write down the three numbers, when:

        i $x = 6$ **ii** $x = 11$ **iii** $x = 2.5$.

    b Write down an expression for the sum of the three numbers.

        Simplify your expression and factorise it as much as possible.

(MR) **9** Three numbers are a, $a + 3$ and $a + 12$.

    a Write down the three numbers, when $a = 10$.

    b Write down the three numbers, when $a = -10$.

    c Write down an expression for the sum of the three numbers.

        Simplify your expression and factorise it as much as possible.

    d Work out the sum of the three numbers, when $a = 45$.

Investigation: Finding factors

A   Look at this expression.

   $24x + 36$

   You could take out 2 as a factor and write:

   $24x + 36 = 2(12x + 18)$

   You could take out 6 as a factor and write:

   $24x + 36 = 6(4x + 6)$

   Write $24x + 36$ in other ways, by taking out different factors.

   How many ways can you find?

B   Look at this expression.

   $36y - 18$

   Write it in as many different ways you can by taking out different factors.

C   Repeat part **B** with the expression $60 + 100z$.

2.3 Equations with brackets

Learning objective

* To solve equations with one or more sets of brackets

You have already learnt how to solve simple equations with brackets.

Example 4

Solve the equation $3(x - 4) = 16$

a by first multiplying out the brackets   **b** by first dividing by 3.

   a $3(x - 4) = 16$

   $3x - 12 = 16$ Multiply x and 4 by 3.

   $3x = 28$ Add 12 to both sides.

   $x = 9\frac{1}{3}$ Divide 28 by 3.

   b $3(x - 4) = 16$

   $x - 4 = 5\frac{1}{3}$ Divide 16 by 3.

   $x = 9\frac{1}{3}$ Add 4 to $5\frac{1}{3}$.

You can use either method to solve equations like these.

If there is more than one set of brackets, it is usually easier to multiply them both out first.

Example 5

Solve the equation $2(x + 1) + 4(x - 3) = 35$.

$$2(x + 1) + 4(x - 3) = 35$$

$$2x + 2 + 4x - 12 = 35 \qquad \text{Multiply out both sets of brackets.}$$

$$6x - 10 = 35 \qquad \text{Collect like terms.}$$

$$6x = 45 \qquad \text{Add 10 to both sides.}$$

$$x = 7.5 \qquad 45 \div 6 = 7\tfrac{1}{2} \text{ or } 7.5$$

Exercise 2C

1 Solve these equations.

Show your method each time.

   a $2x + 6 = 18$ **b** $2(x + 6) = 18$ **c** $4y - 2 = 20$ **d** $4(y - 2) = 20$

2 Solve these equations. Show your method.

   a $2(y - 8) = 20$ **b** $2(y - 8) = 10$ **c** $4(y + 2) = 32$ **d** $3(y + 4) = 12$

   e $4(3 + t) = 48$ **f** $3(6 + k) = 27$ **g** $40 = 5(f - 17)$ **h** $20 = 4(w - 9)$

3 Solve these equations. Give your answers in fraction form.

   a $3(x - 2) = 5$ **b** $4(y + 1) = 11$ **c** $6(t - 4) = 19$ **d** $8(c + 3) = 25$

4 **a** Simplify the expression $4(x + 2) + x - 5$.

   b Solve the equation $4(x + 2) + x - 5 = 38$.

5 **a** Simplify the expression $2(x - 1) + 6(x - 2)$.

   b Solve the equation $2(x - 1) + 6(x - 2) = 30$.

6 **a** Show that the expression $3(x + 4) - 2(x + 2)$ simplifies to $x + 8$.

   b Solve the equation $3(x + 4) - 2(x + 2) = 20$.

   c Solve the equation $3(x + 4) - 2(x + 2) = 9$.

7 **a** Show that $4(x + 1) - 2(x - 3)$ simplifies to $2x + 10$.

   b Solve the equation $4(x + 1) - 2(x - 3) = 24$.

   c Solve the equation $4(x + 1) - 2(x - 3) = 13$.

8 Solve these equations.

   a $2(x + 3) + 2(x + 4) = 28$ **b** $3(x - 3) + 2(x + 4) = 34$ **c** $4(y - 1) - 2(y + 2) = 17$

   d $3(x - 4) - 2(x + 5) = 0$ **e** $5(d + 3) - 2(d - 1) = 29$ **f** $4(t - 3) - (t - 5) = 20$

9 Solve each of these equations.

Start by multiplying out the brackets.

   a $5(x - 1) = 4(x + 1)$ **b** $3(a - 2) = 2(a + 2)$ **c** $5(t - 3) = 3(t + 5)$

   d $3(x + 1) = 2(x + 3)$ **e** $4(x - 2) = 5(x - 3)$ **f** $6(x - 1) = 4(x + 5)$

   2 Equations and formulae

Challenge: Six cards

The lengths of this card are in centimetres.

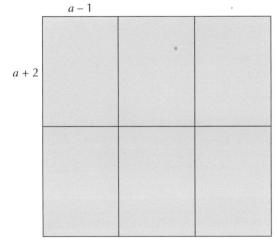

A Work out the length of the perimeter of the card, when:

   a $a = 4$ **b** $a = 6$.

Here are six identical cards arranged in a rectangle.

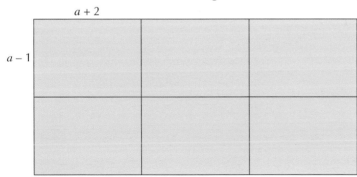

B Work out an expression for the length of the perimeter of this rectangle.

   Simplify it as much as possible.

C Here are the same six cards arranged in a different rectangle.

Show that the perimeter of this rectangle is 6 cm longer than the perimeter of the previous rectangle.

2.4 Equations with fractions

Learning objective

• To solve equations involving fractions

If there is a fraction in an equation you can remove it by multiplying the whole equation by the denominator of the fraction.

Example 6

Solve the equation $\frac{3}{5}x = 8$.

$\frac{3}{5}x = 8$

$3x = 40$ Multiply both sides by 5. $8 \times 5 = 40$

$x = 13\frac{1}{3}$ $40 \div 3 = 13$ remainder 1

In the example, it is better to leave the answer as a fraction because you cannot write it exactly as a decimal.

Example 7

Solve the equation $\frac{2}{3}(a - 4) = 5$

$\frac{2}{3}(a - 4) = 5$

$2(a - 4) = 15$ Multiply both sides by 3.

                     15 is 3×5 and you now have an integer in front of the bracketed term.

$a - 4 = 7.5$ Divide by 2.

$a = 11.5$ Add 4.

Equations with fractions can be written in different ways. Look at the equations in the examples.

$\frac{3}{5}x = 8$ could be written as $\frac{3x}{5} = 8$.

$\frac{2}{3}(a - 4) = 5$ could be written as $\frac{2(a - 4)}{3} = 5$.

Example 8

Solve the equation $\frac{12}{x + 1} = 5$.

$\frac{12}{x + 1} = 5$ $x + 1$ is in the denominator so multiply by it to remove the fraction.

$12 = 5(x + 1)$ Solve this in the usual way. Expand the term with brackets first.

$12 = 5x + 5$ Now subtract 5.

$7 = 5x$ Then divide by 5.

$x = 1\frac{2}{5}$ or 1.4

Exercise 2D

1 Solve these equations.

a $\frac{x}{5} = 6$ b $\frac{y}{3} = 7$ c $\frac{t}{4} = 4$ d $\frac{n}{2} = 15$ e $\frac{m}{15} = 2$

f $\frac{1}{3}x = 2$ g $\frac{1}{2}z = 5$ h $\frac{1}{4}t = 10$ i $\frac{1}{8}v = 3$ j $\frac{1}{5}h = 5$

2 Solve these equations.

a $\frac{2}{3}x = 8$ b $\frac{2}{3}x = 12$ c $\frac{2}{3}x = 20$ d $\frac{2}{3}x = 3$ e $\frac{2}{3}x = 7$

3 Solve these equations.

a $\frac{1}{5}x = 4$ b $\frac{2}{5}x = 4$ c $\frac{3}{5}x = 4$ d $\frac{4}{5}x = 4$

4 Solve these equations.

a $\frac{20}{x} = 4$ b $\frac{20}{x} = 10$ c $\frac{15}{y} = 6$ d $\frac{30}{t} = 5$

e $\frac{100}{n} = 5$ f $\frac{19}{m} = 4$ g $\frac{5}{x} = 10$ h $\frac{4}{y} = 20$

5 Solve these equations.

a $\frac{3}{4}a = 6$ b $\frac{3}{8}b = 4$ c $\frac{5}{8}c = 2$ d $\frac{2}{7}k = 4$ e $\frac{3}{5}t = 9$

f $\frac{2c}{3} = 8$ g $\frac{3m}{5} = 2$ h $\frac{4t}{5} = 6$ i $\frac{5f}{8} = 4$ j $\frac{2k}{5} = 30$

6 Solve these equations.

a $\frac{1}{4}(x + 2) = 5$ b $\frac{1}{3}(x - 4) = 2$ c $\frac{1}{6}(x + 13) = 4$ d $\frac{1}{8}(x - 9) = 3$

e $\frac{t + 3}{2} = 6$ f $\frac{x - 5}{3} = 4$ g $\frac{n - 10}{6} = 3$ h $\frac{a + 7}{4} = 9$

(MR) 7 The width of this rectangle is 6 cm and the area is $2n + 3$ cm².

6 cm | 2n + 3 cm²

a Explain why the length of the rectangle is $\frac{2n + 3}{6}$ cm.

b You are told that the length of the rectangle is 9 cm.

Write down an equation and solve it to work out the value of n.

MR **8** Four numbers are x, $x + 4$, $x - 3$, and $2x$.

    a Work out the values of the four numbers, given that $x = 12$.

    b Work out the values of the four numbers, given that $x = 5\frac{1}{2}$.

    c Show that the mean of the four numbers is $\dfrac{5x + 1}{4}$.

    d Suppose the mean of the four numbers is 11.

        Write down an equation and solve it to work out the value of x.

    e Work out the four numbers for the value of x you found in part **d**.

9 Solve these equations.

    a $\dfrac{16}{x + 2} = 8$ **b** $\dfrac{48}{y - 1} = 8$ **c** $\dfrac{30}{k - 5} = 5$ **d** $\dfrac{36}{d + 3} = 4$

10 Solve these equations.

    a $\frac{2}{3}(x + 2) = 8$ **b** $\frac{3}{4}(t - 1) = 2$ **c** $\frac{2}{3}(z + 5) = 7$ **d** $\frac{3}{5}(r - 2) = 4$

    e $\frac{2}{5}(a + 3) = 5$ **f** $\frac{3}{8}(t + 10) = 6$ **g** $\frac{3}{8}(z - 2) = 4$ **h** $\frac{3}{8}(m - 11) = 2$

11 Solve these equations.

    a $\dfrac{2(x + 5)}{3} = 16$ **b** $\dfrac{3(t - 2)}{4} = 4$ **c** $\dfrac{3(2 + x)}{10} = 4$ **d** $\dfrac{5(c - 10)}{9} = 2$

    e $\dfrac{6(r - 8)}{5} = 2$ **f** $\dfrac{2(3 + t)}{5} = 4$ **g** $\dfrac{9(d - 2)}{10} = 2$ **h** $\dfrac{3(e - 5)}{4} = 4$

Mathematical reasoning: Making equations

Hassan makes up an equation like this.

    Start with the answer of 7. $x = 7$

    Add 3 to both sides. $x + 3 = 10$

    Multiply both sides by 2. $2(x + 3) = 20$

    Divide both sides by 5. $\dfrac{2(x + 3)}{5} = 4$

A Solve Hassan's equation in the usual way, to check that the solution is $x = 7$.

B **a** Make up an equation by following these steps.

      Start with an answer of 18.

      Subtract 6 from both sides.

      Multiply both sides by 5.

      Divide both sides by 4.

   b Solve your equation in part **a** to check it gives the correct answer.

C Use the same method to make up an equation of your own.

    Give it to someone else to solve.

2.5 Rearranging formulae

Learning objective

• To change the subject of a formula

Key words

rearrange

subject

Look at this simple formula.

$A = B + 2$

Here A is the **subject** of the formula.

If you subtract 2 from both sides you get: $\quad A - 2 = B$

You can write this the other way round as: $\quad B = A - 2$

Now you have **rearranged** the formula to make B the subject.

Example 9

This is the formula for the perimeter of a rectangle.

$p = 2(l + w)$

Rearrange it to make w the subject.

$p = 2(l + w)$ You need to get w on its own on one side of the equation.

$\dfrac{p}{2} = l + w$ First divide by 2. You can write either $\dfrac{1}{2}p$ or $\dfrac{p}{2}$ on the left-hand side.

$\dfrac{p}{2} - l = w$ Subtract l from both sides.

$w = \dfrac{p}{2} - l$ You should usually put the subject on the left-hand side.

Exercise 2E

1   Rearrange each formula to make v the subject.

    a $t = v + 5$ **b** $t = \frac{1}{2}(v + 5)$ **c** $t = 2v + 5$ **d** $t = 2(v + 5)$

2   These are the equations of some lines. Rearrange each one to make x the subject.

    a $y = x - 5$ **b** $y = 2x$ **c** $y = 3x + 6$ **d** $y = \frac{1}{2}(x + 3)$

    e $x + y = 10$ **f** $y = 4(x - 1)$ **g** $y = \frac{1}{2}(x + 7)$ **h** $y + 2x = 9$

3   This is a formula connecting A, b and h in the triangle.

    $2A = bh$

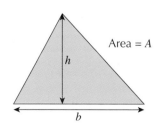

Area = A

    a   Rearrange the formula to make A the subject.

    b   Rearrange the formula to make b the subject.

    c   Rearrange the formula to make h the subject.

    d   Work out the value of A when $b = 12$ and $h = 9$.

    e   Work out the value of b when $A = 100$ and $h = 16$.

    f   Work out the value of h when $A = 8.64$ and $b = 3.6$.

4 This shape is a hexagon.

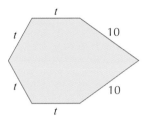

a Show that the formula for the perimeter, p, of this hexagon is $p = 4t + 20$.

b Rearrange the formula to make t the subject.

c Use your formula from part **b** to find the value of t when $p = 50$.

5 This is a formula used in science.

$$I = \frac{V}{R}$$

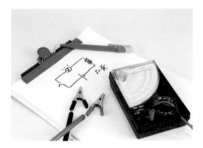

a Find the value of I when $V = 9$ and $R = 18$.

b Rearrange the formula to make V the subject.

c Rearrange the formula to make R the subject.

6 This is another formula that is used in science.

$$s = \tfrac{1}{2}(u + v)t$$

a Work out the value of s when $u = 2.5$, $v = 5.5$ and $t = 12$.

b Show that the formula can be rearranged as:

$$t = \frac{2s}{u + v}$$

c Use the formula in part **b** to work out the value of t when $s = 40$, $u = 3$ and $v = 7$.

7 Here are four formulae.

$$ab = c - d \qquad a = \frac{c - d}{b} \qquad c = ab + d \qquad d = ab - c$$

Which is the odd one out? Justify your answer.

8 A formula for the area, A, of this triangle is $A = \frac{1}{2}xy$.

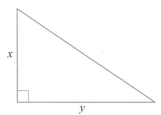

Copy and complete these rearrangements of the formula.

a $xy = \ldots$ **b** $y = \ldots$ **c** $x = \ldots$

9 This is a formula used in science.

$$k = \frac{1}{2}mv^2$$

Copy and complete these rearrangements of the formula.

a $mv^2 = \ldots$ **b** $v^2 = \ldots$ **c** $m = \ldots$

Challenge: Equation of a line

The equation of this line is $2x + 3y = 60$.

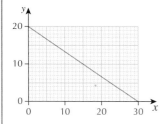

A Show, by substitution, that the point $(13, 11\frac{1}{3})$ is on this line.

B Rearrange the formula to make x the subject.

C The point $(x, 7)$ is on the line. Use your formula from question **B** to find the value of x at this point.

D Rearrange the formula to make y the subject.

E The point $(10\frac{1}{2}, y)$ is on the line. Use your formula from question **D** to find the value of y at this point.

F These points are on the line. Use the appropriate formula to find the missing values.

$(a, 2)$ $(19, b)$ $(c, 9)$ $(17\frac{1}{2}, d)$

Ready to progress?

  I can expand brackets.

  I can factorise algebraic expressions.
I can solve equations that have brackets or fractions or both.

  I can rearrange simple formulae.

Review questions

1   Multiply out these brackets.

   a $4(t - 3)$ **b** $3(a + b - 6)$ **c** $5(3 - x + y)$

2   Multiply out these brackets.

   a $2(3x + y)$ **b** $4(2a - 4c)$ **c** $4(3k - 5)$ **d** $2(4 - 3r + 2s)$

3   Simplify these expressions as much as possible. Multiply out the brackets first.

   a $2(a + 3) + 4$ **b** $5(2t - 3) - 7$ **c** $4n + 3(n - 2)$ **d** $4n - 3(n - 4)$

4   Simplify these expressions as much as possible. Multiply out the brackets first.

   a $4(a + b) + 2(a + b)$ **b** $4(x + 5) - 2(x + 3)$ **c** $2(t - 3) + 3(t - 2)$
   d $2(x + 2) - 2(x + 1)$ **e** $4(d + 2) - 2(d + 4)$ **f** $2y + 2(y + 1) - 2(y - 1)$

5   Solve these equations.

   a $\frac{3}{4}x = 12$ **b** $\frac{2}{5}x = 5$ **c** $\frac{3x}{8} = 6$ **d** $\frac{4x}{7} = 3$

6   Solve these equations.

   a $2(x + 5) = 9$ **b** $\frac{x + 5}{3} = 9$ **c** $\frac{2}{3}(x + 5) = 9$ **d** $\frac{2}{9}(x + 5) = 3$

7   **a**   You are given that $x = 3$. Work out the value of $2(x + 4) + 3(x - 2)$.

   b   Now solve the equation $2(x + 4) + 3(x - 2) = 42$.

   c   Solve the equation $2(x + 4) + 3(x - 2) = 30$.

8   Solve these equations.

   a $4(t + 4) + 6(t - 3) = 40$ **b** $5(d + 1) + 3(d - 1) = 20$

   c $6(n - 4) = 3(n + 15)$ **d** $2(9 - x) = 3(1 + x)$

9 Here is a formula.

$$F = 3.4G + 2.5H$$

a   Make G the subject of the formula.

b   Make H the subject of the formula.

 10 Joe invested £1000 for T years and the rate of simple interest was $R\%$.

The amount of interest, £I, is given by the formula $I = 10RT$.

a   Work out the value of I when $T = 2$ and $R = 4.5$.

b   Suppose $I = 75$ and $T = 3$. Work out the value of R.

c   Make R the subject of the formula.

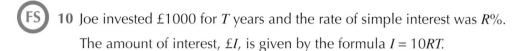 11 This is a formula used in science.

$$s = \tfrac{1}{2}(u + v)t$$

a   Show that the formula can be rewritten as $2s = ut + vt$.

b   Rewrite the formula to make v the subject.

Investigation
Body mass index

Body mass index (BMI) is a simple measure for human body shape. It is based on mass, M, in kilograms and height, H, in metres.

The formula for body mass index (I) is: $I = \frac{M}{H^2}$

For example Lennox Lewis, a former heavyweight boxer, is 1.96 m tall and his mass is 93 kg.

Lennox Lewis' body mass index is $\frac{93}{1.96^2} = 24.2$.

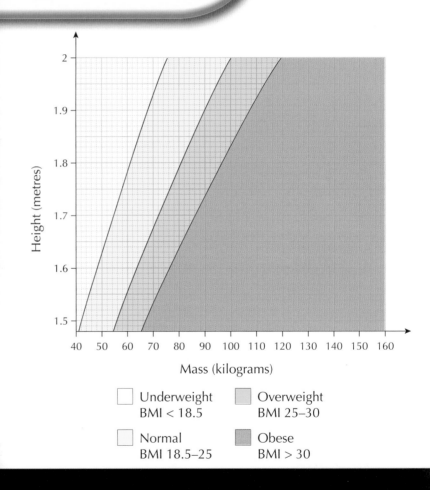

Underweight
BMI < 18.5

Overweight
BMI 25–30

Normal
BMI 18.5–25

Obese
BMI > 30

2 Equations and formulae

1 Use a calculator to check that the value given for Lennox Lewis is correct, to one decimal place. For people with a fairly active lifestyle, a value between 18.5 and 25 is normal. A value outside this range may indicate that a person is overweight or underweight. However, this is only a rough guide and there are other reasons why a person may be outside this range, such as body fat and muscularity. These categories are only used for adults, not for children.

2 This table shows the masses and heights of some well-known people. Work out the BMI of each one.

Name	Profession	Mass (kg)	Height (m)
Brad Pitt	Actor	78	1.80
Angelina Jolie	Actress	59	1.73
Arnold Schwarzenegger	Actor/politician	113	1.88
Serena Williams	Tennis player	68	1.75
Ben Morgan	Rugby player	116	1.91
Kate Moss	Model	55	1.70

3 Check where each person in the table would be in the chart. Decide whether, according to the chart, each one is in the normal range, underweight, overweight or obese.

4 Japanese Sumo wrestlers are very fit but they deliberately build up their body mass as much as they can. A recent champion is Hakuho Sho, who is 1.93 m tall and has a mass of 155 kg. What is his BMI? How would he be classified, according to the chart?

5 The formula you are using is $I = \frac{M}{H^2}$

Rearrange the formula to make M the subject.

6 These are the BMIs for some well-known people. Use your formula from question 4 to calculate the mass of each person.

Name	Profession	BMI	Height (m)
David Beckham	Footballer	22.4	1.83
Beyoncé Knowles	Singer	22.4	1.69
Bradley Wiggins	Cyclist	19.1	1.90
Jessica Ennis	Athlete	23.1	1.57
Usain Bolt	Athlete	22.3	1.96
Claudia Schiffer	Model	17.8	1.80

3

Polygons

This chapter is going to show you:

- how to calculate the interior and exterior angles of polygons
- how to calculate the interior and exterior angles of regular polygons
- how regular polygons tessellate.
- how to make accurate geometric constructions

You should already know:

- the names of polygons
- the sum of the interior angles of a triangle and of a quadrilateral
- how to solve equations.

About this chapter

Flanked by the wild North Atlantic Ocean on one side and a landscape of dramatic cliffs on the other, for centuries the Giant's Causeway has inspired artists, stirred scientific debate and captured the imagination of all who see it. Storytellers have their own explanation for this captivating stretch of coast, and many tales endure to the present day. The most famous legend associated with the Giant's Causeway is that of an Irish giant, Finn McCool. The causeway was believed to be the remains of the bridge that Finn built, to link Ireland to Scotland. The landscape became so imbued with the spirit of this legend that it gave rise to the name – the Giant's Causeway.

The formation consists of about 40 000 interlocking basalt columns, most of which are hexagonal. The columns form huge stepping stones, some as high as 39 feet (about 12 metres), which slope down to the sea. Some of the columns have four, seven or eight sides. Weathering of the rock formation has also created circular structures, which the locals call 'giant's eyes'.

The Giant's Causeway was actually formed by intense volcanic activity, about 50 million years ago. As the lava rapidly cooled, it contracted into the distinctive shapes.

3.1 Angles in polygons

Learning objectives

- To work out the sum of the interior angles of a polygon
- To work out exterior angles of polygons

Key words	
exterior angle	interior angle
irregular polygon	polygon

A **polygon** is a 2D shape that has straight sides.

The angles inside a polygon are called **interior angles**.

You already know that:

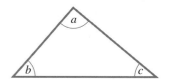

- for any triangle, the sum of the interior angles, a, b and c, is 180°

  $a + b + c = 180°$

- for any quadrilateral, the sum of the interior angles, a, b, c and d, is 360°.

  $a + b + c + d = 360°$

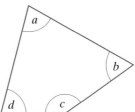

These are some examples of **irregular polygons**.

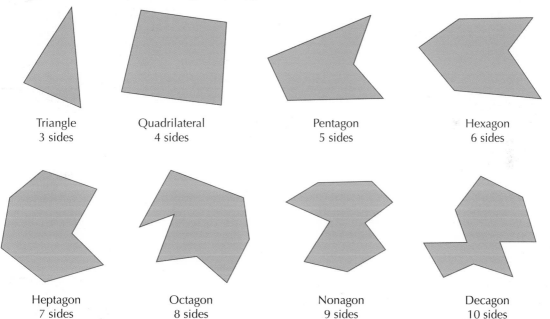

| Triangle | Quadrilateral | Pentagon | Hexagon |
| 3 sides | 4 sides | 5 sides | 6 sides |

| Heptagon | Octagon | Nonagon | Decagon |
| 7 sides | 8 sides | 9 sides | 10 sides |

Example 1

Work out the sum of the interior angles of a pentagon.

The diagram shows how a pentagon can be split into three triangles from one of its vertices. The sum of the interior angles for each triangle is 180°.

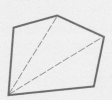

So, the sum of the interior angles of a pentagon is given by:

$3 × 180° = 540°$

If a side of a polygon is extended, the angle formed outside the polygon is an **exterior angle**.

In the diagram, a is an exterior angle of the quadrilateral.

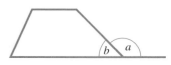

At any vertex of a polygon:

the interior angle plus the exterior angle = 180° (angles on a straight line)

Hence, on the diagram:

$a + b = 180°$

 Hint This has been shown for a quadrilateral, but remember that it is true for any polygon.

Example 2

What can you say about the exterior angles of a pentagon?

In the diagram, each side of the pentagon has been extended to show all the exterior angles.

Imagine standing on each vertex of the pentagon, in turn.

Start at the fist vertex, facing along the extended line.

Turn and walk along the other line of the angle, as far as the second vertex.

Turn, and walk along the next line.

As you go round the pentagon, you turn clockwise through all of its exterior angles in turn.

You can see that you will have turned through 360°.

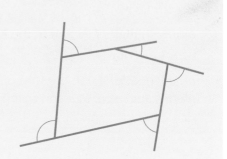

This is true for all polygons.

The sum of the exterior angles for any polygon is 360°.

Exercise 3A

 1 **a** Sketch each polygon. By splitting it into triangles, find the sum of the interior angles of: **i** a hexagon   **ii** an octagon.

   b Copy and complete this table.
   The first three have been done for you.

   Hint Look for patterns in the table.

   Do not draw the polygons.

Name of polygon	Number of sides	Number of triangles inside polygon	Sum of interior angles
Triangle	3	1	180°
Quadrilateral	4	2	360°
Pentagon	5	3	540°
Hexagon			
Heptagon			
Octagon			
Nonagon			
Decagon			

2 Calculate the unknown angle in each polygon.

a

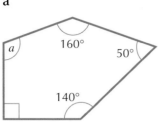

160°
50°
a
140°

b

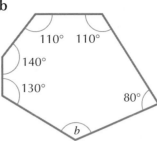

110° 110°
140°
130°
80°
b

c

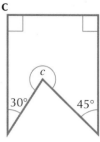

c
30° 45°

3 Calculate the unknown angle in each diagram.

a

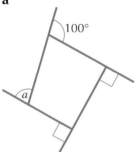

100°
a

b

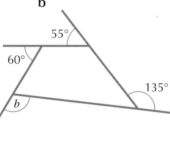

55°
60°
135°
b

c

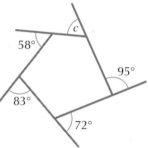

c
58°
95°
83°
72°

4 Calculate the value of *x* in this pentagon.

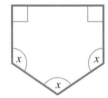

x   *x*
x

5 The four interior angles of a quadrilateral are $3x + 80°$, $5x + 10°$, $3x − 20°$ and $4x − 10°$.
Calculate the size of each interior angle of the quadrilateral.

MR **6** Calculate the size of each unknown angle.
For each angle explain, with a reason, how you found the value.

a

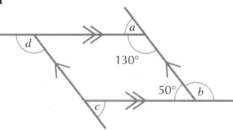

a
d
130°
50° *b*
c

b

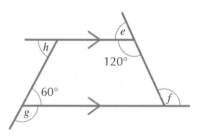

e
h
120°
60°
g
f

For each diagram, check that the sum of the exterior angles is 360°.

7 A triangle has no diagonals, a quadrilateral has two diagonals and a pentagon has five diagonals.

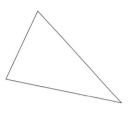 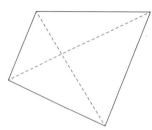 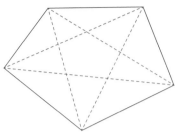

a Show that a hexagon has nine diagonals.

b How many diagonals does a heptagon have?

c Work out how many diagonals an octagon has.

d This is a formula that connects the number of diagonals, d, to the number of sides, s, for any polygon.

$$d = \frac{s(s - \ldots)}{\ldots}$$

Two numbers are missing from the formula.

Work out the two missing numbers.

> **Hint** The two numbers are not the same.

e Use the formula to work out the number of diagonals in:

i a nonagon   **ii** a decagon.

Activity: Constructing triangles

Construct the triangle PQR accurately.

- Draw the line QR, 6 cm long.

- Set the compasses to a radius of 4 cm and, with centre at Q, draw a long arc above QR.

- Set the compasses to a radius of 5 cm and, with centre at R, draw a long arc to intersect the first arc.

- The intersection of the arcs is the point P.

- Join QP and RP to complete the triangle.

This is an example of constructing a triangle, given three sides (SSS).

You should leave the two construction lines on the diagram.

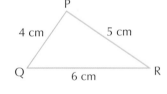

Construct these triangles accurately.

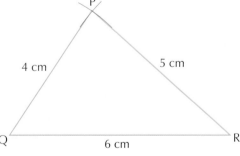

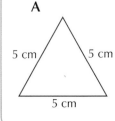 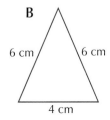 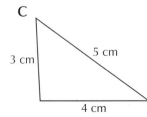 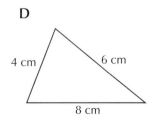

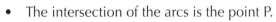

3.2 Constructions

Learning outcome

- To make accurate geometric constructions

An important part of your study of geometry is learning how to make accurate constructions.

The next four examples will show you some important geometric constructions. They are useful because they produce exact measurements and are therefore used by architects and also in design and technology.

You will need a sharp pencil, a ruler and compasses.

Always leave all your construction lines on your diagrams.

Example 3

Construct the perpendicular from point P to the line segment AB.

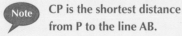

- Set the compasses to any suitable radius.
  Draw arcs from P to intersect AB at X and Y.
- With the compasses still set at the same radius, draw arcs centred on X and Y to intersect at Z below AB.
- Join PZ.
- PZ is perpendicular to AB and intersects AB at a point C.

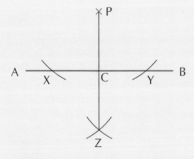

> **Note** CP is the shortest distance from P to the line AB.

Example 4

Construct the perpendicular at point Q on the line segment XY.

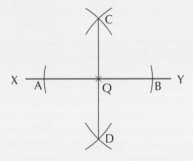

- Set the compasses to a radius that is less than half the length of XY.

  With the centre at Q, draw arcs on either side of Q to intersect XY at A and B.
- Set the compasses to a radius that is greater than half the length of XY and, with centre at A and then B, draw arcs above and below XY to intersect at C and D.
- Join CD.
- CD is the perpendicular from the point Q.

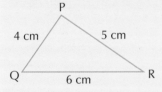

Example 5

Construct the triangle PQR.

- Draw the line QR 6 cm long.
- Set the compasses to a radius of 4 cm and, with centre at Q, draw a long arc above QR.
- Set the compasses to a radius of 5 cm and, with centre at R, draw a long arc to intersect the first arc.
- The intersection of the arcs is the point P.
- Join QP and RP to complete the triangle.

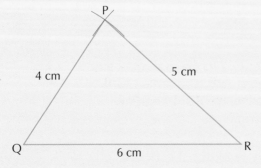

Note This is an example of constructing a triangle, given three sides (SSS).

Example 6

Construct right-angled triangle ABC.

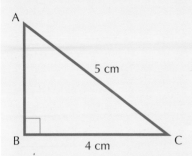

- Draw the line BC 4 cm long.

- Use the method in Example 4 to construct the perpendicular at B.

  (You will need to extend the line CB.)

- Set your compasses to a radius of 5 cm and with the centre at C, draw an arc to intersect the perpendicular from B.

- The intersection of the arc and the perpendicular is the point A.

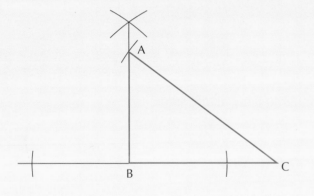

Note This is an example of constructing a triangle, given the longest side (called the hypotenuse), a shorter side and a right angle (RHS).

Exercise 3B

1 Copy the diagram and construct the perpendicular from the point X to the line AB.

X x

A ———————————— B

2 Draw a circle of radius 6 cm and centre O. Draw a line AB of any length across the circle, as in the diagram. (AB is called a chord.) Construct the perpendicular from O to the line AB. Extend the perpendicular to make a diameter of the circle.

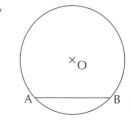

3 a Copy the diagram and construct the perpendicular from the point Z on the line segment XY.

b Copy the diagram and construct the perpendicular from the point C on the line segment AB.

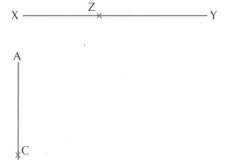

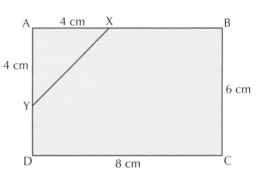

4 ABCD is a rectangle. X and Y are points on AB and AD respectively.

Use a suitable construction to work out the shortest distance from C to XY.

5 Construct each triangle. Remember to label all the sides.

a

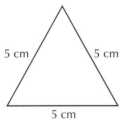

b

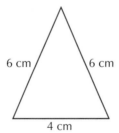

c

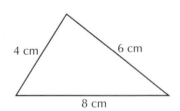

6 Construct each right-angled triangle. Remember to label all the sides.

a

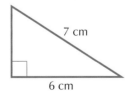

b

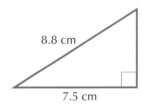

c

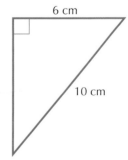

3.3 Angles in regular polygons

Learning objectives

- To work out the exterior angles of a regular polygon
- To work out the interior angles of a regular polygon

Key word

regular polygon

In a **regular polygon**, all the interior angles are equal and all the sides have the same length.

These are some examples of regular polygons.

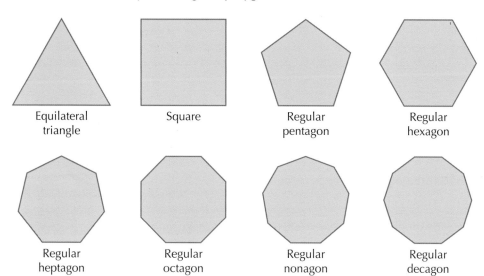

| Equilateral triangle | Square | Regular pentagon | Regular hexagon |

| Regular heptagon | Regular octagon | Regular nonagon | Regular decagon |

Example 7

Work out the size of each exterior and each interior angle in a regular pentagon.

A regular pentagon has five equal exterior angles.

Let the size of each exterior angle be x, as shown on the diagram.

The sum of all the exterior angles is 360°. This gives:

$$5x = 360°$$

$$x = \frac{360°}{5} = 72°$$

The regular pentagon has five equal interior angles.

Let the size of each interior angle be y, as shown on the diagram.

The sum of an interior angle and an exterior angle is 180°. This gives:

$$y + 72° = 180°$$

$$y = 180° - 72° = 108°$$

1   Copy and complete the table below for regular polygons.
    The first three rows have been done for you.

Regular polygon	Number of sides	Sum of exterior angles	Size of each exterior angle	Size of each interior angle
Equilateral triangle	3	360°	120°	60°
Square	4	360°	90°	90°
Regular pentagon	5	360°	72°	108°
Regular hexagon				
Regular octagon				
Regular nonagon				
Regular decagon				

2   A regular dodecagon is a polygon with 12 sides.
    All 12 interior angles are equal and all 12 sides have the same length.
    a   Work out the size of each exterior angle.
    b   Work out the size of each interior angle.
    c   Calculate the sum of the interior angles.

3   ABCDE is a regular pentagon.
    a   What type of triangle is ADE?
    b   Calculate the size of angle DAE.
    c   Hence calculate the size of angle CAD.

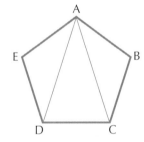

4   ABCDE is a regular pentagon. The sides BC and ED
    are extended to meet at X.
    Calculate the size of angle CXD.

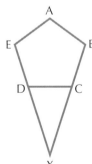

5   Two lines of symmetry are drawn on a regular hexagon,
    as shown on the diagram.
    Write down the size of each of the angles a, b and c.

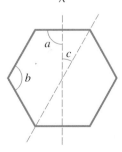

PS **6** ABCDEFGH is a regular octagon.

Copy the diagram and explain how to calculate the size of angle AFB.

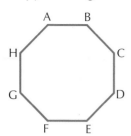

MR **7** Each exterior angle of a regular polygon is 20°.

How many sides does the polygon have?

MR **8** Each interior angle of a regular polygon is 162°.

How many sides does the polygon have?

Challenge: Exterior and interior angles as mixed numbers

The size of exterior and interior angles for some regular polygons are not whole numbers of degrees.

Work out the size of each exterior angle and each interior angle in:

A a regular heptagon

B a regular hendecagon (an 11-sided polygon)

C a regular tridecagon (a 13-sided polygon).

Give your answers as mixed numbers.

Hint Use the $a\,{}^{b}/_{c}$ key or the ▢ key on your calculator, to work out fractions.

3.4 Regular polygons and tessellations

Learning objective

* To work out which regular polygons tessellate

Key words	
semi-tessellation	tessellate
tessellation	

A **tessellation** is a repeating pattern made from identical 2D shapes that fit together exactly, leaving no gaps.

This section will show you which of the regular polygons **tessellate**.

Hint To show how a shape tessellates, draw up to about ten repeating shapes.

Example 4

Draw diagrams to show how equilateral triangles and squares tessellate.

Equilateral triangles tessellate like this.

Squares tessellate like this.

Exercise 3D

1 Show how a regular hexagon tessellates. Use an isometric grid.

2 Trace this regular pentagon onto card and cut it out to make a template.

   a Use your template to show that a regular pentagon does not tessellate.

   b Explain why a regular pentagon does not tessellate.

3 Trace this regular octagon onto card and cut it out to make a template.

   a Use your template to show that a regular octagon does not tessellate.

   b Explain why a regular octagon does not tessellate.

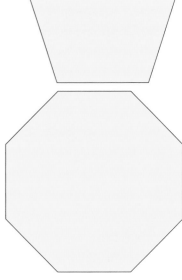

(MR) 4 **a** Copy and complete the table below for regular polygons.

   b Use your table to explain why only some of the regular polygons tessellate.

   c Do you think that a regular nonagon tessellates? Explain your answer.

Regular polygon	Size of each interior angle	Does this polygon tessellate?
Equilateral triangle		
Square		
Regular pentagon		
Regular hexagon		
Regular octagon		

5 Polygons can be combined to form a **semi-tessellation**.

Here are two examples.

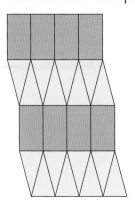 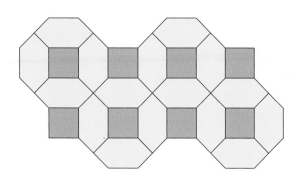

Rectangles and isosceles triangles   Squares and hexagons

Invent your own semi-tessellations. Use squared paper or isometric paper.

Activity: Regular semi-tessellations

Work in pairs or small groups.

Regular polygons can be combined together to form regular semi-tessellations.

Here is an example.

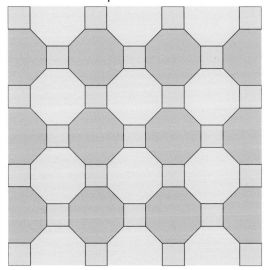

This is made from squares and regular octagons.

A Draw some other regular semi-tessellations and make a poster to display them in your classroom.

B There are eight regular semi-tessellations altogether. Try to find them all.

Ready to progress?

I know the sum of the interior angles of a triangle and of a quadrilateral.
I know how to tessellate 2D shapes.

I can work out and use interior and exterior angles of polygons.
I can work out and use interior and exterior angles of regular polygons.

Review questions

1   a   Explain why the sum of the interior angles in a quadrilateral is 360°.

b   What is the sum of the interior angles in a pentagon?

c   What is the sum of the interior angles in a heptagon?

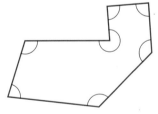

2   The diagram shows a rectangle that just touches an equilateral triangle.

a   Work out the size of the angle marked *x*.

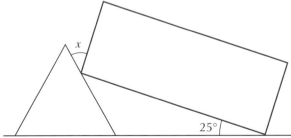

b Now the rectangle just touches the equilateral triangle, so that ABC is a straight line.

Show that triangle BDE is isosceles.

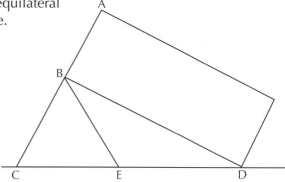

3 Calculate the size of angle *a* in the hexagon.

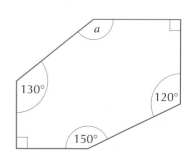

4 The five exterior angles of a pentagon are $2x$, $3x + 10°$, $4x + 20°$, $3x - 50°$ and $80° - 2x$.

Work out the size of each interior angle of the pentagon.

5 ABCDEF is a regular hexagon. The two lines of symmetry AD and BE intersect at X.

   a Calculate the size of angle AXB.

   b What type of triangle is triangle AXB?

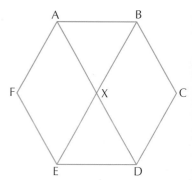

6 The diagram shows a five-pointed star or pentagram.

It is made from a regular pentagon and five isosceles triangles.

Calculate the value of *x*.

7 This pattern is made from tessellating squares.

What percentage of the pattern is shaded red?

Activity
Garden design

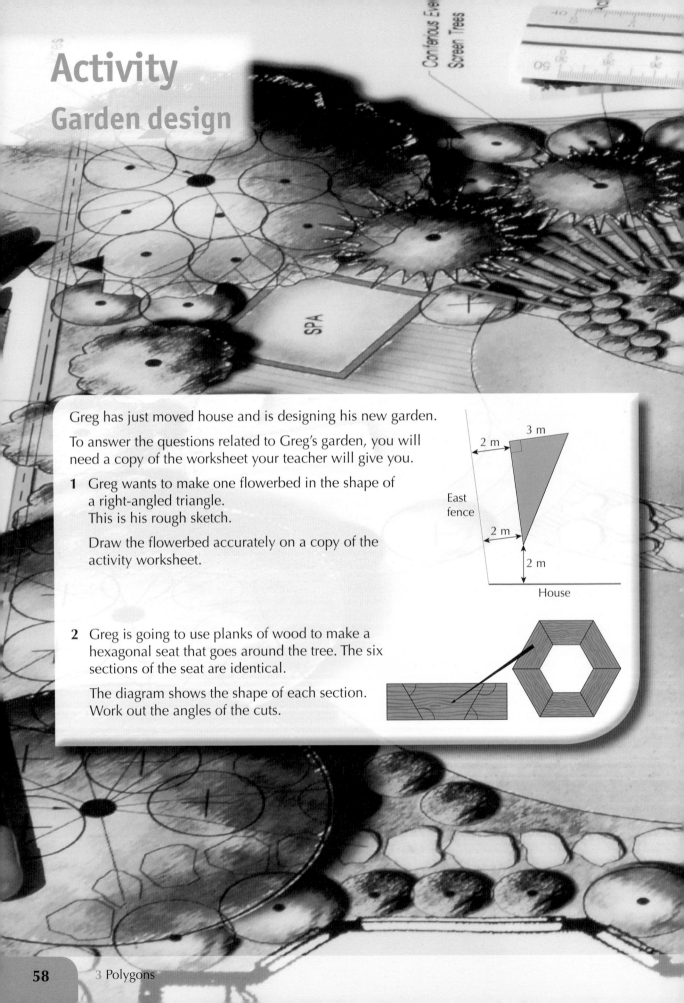

Greg has just moved house and is designing his new garden.

To answer the questions related to Greg's garden, you will need a copy of the worksheet your teacher will give you.

1 Greg wants to make one flowerbed in the shape of a right-angled triangle.
This is his rough sketch.

Draw the flowerbed accurately on a copy of the activity worksheet.

2 Greg is going to use planks of wood to make a hexagonal seat that goes around the tree. The six sections of the seat are identical.

The diagram shows the shape of each section. Work out the angles of the cuts.

3 Greg is planning to lay gravel on the ground, on an area up to 2 m away from the centre of the tree. He will use a length of thick rope to stop the gravel spilling onto the lawn.

Draw an arc on your worksheet, to show where the rope will go.

4 There will be a path, 1 metre wide, around the BBQ area. The path will be made from square paving slabs measuring 50 cm by 50 cm.

   a Draw the path on the activity worksheet.

   b What is the area of the path?

   c How many paving slabs will Greg need?

5 In front of the BBQ area, there will be a large semi-circular table, as shown here.

Draw the table accurately on the activity worksheet.

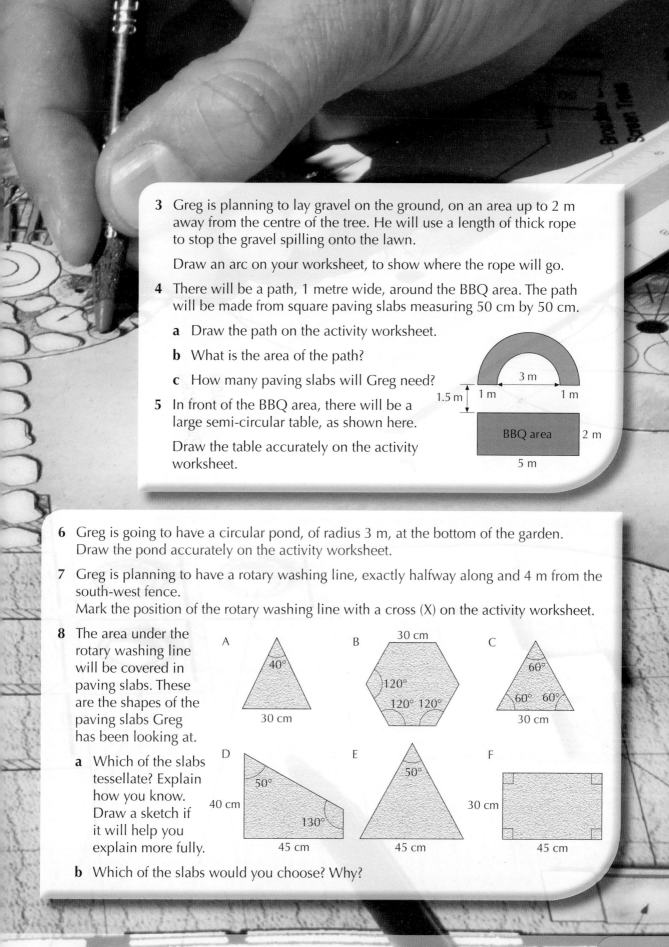

6 Greg is going to have a circular pond, of radius 3 m, at the bottom of the garden. Draw the pond accurately on the activity worksheet.

7 Greg is planning to have a rotary washing line, exactly halfway along and 4 m from the south-west fence.
Mark the position of the rotary washing line with a cross (X) on the activity worksheet.

8 The area under the rotary washing line will be covered in paving slabs. These are the shapes of the paving slabs Greg has been looking at.

   a Which of the slabs tessellate? Explain how you know. Draw a sketch if it will help you explain more fully.

   b Which of the slabs would you choose? Why?

4

Using data

This chapter is going to show you:

- how to interpret correlation from two scatter graphs
- how to interpret time-series graphs
- how to construct and interpret two-way tables
- how to compare two sets of data from statistical diagrams
- how to plan a statistical investigation.

You should already know:

- how to calculate averages – mean, median and mode
- how to use a suitable method to collect data
- how to draw and interpret graphs for discrete data
- how to use mode, median, mean and range to compare two sets of data.

About this chapter

When the weather is hot, many people go to the beach. Sales of sunscreen, ice cream, swimwear and sunshades are high. What happens when the weather is cold?

It is easy to see that some things are related. People like to keep cool and protect themselves from the hot sunshine.

However, it is not always clear whether other information is related in the same way. Do tall people have large handspans? Do cars go faster the bigger their wheels are?

In this chapter, you will find out how to compare two sets of data, to find out if they are related to each other.

4.1 Scatter graphs and correlation

Learning objective

• To infer a correlation from two related scatter graphs

The maximum temperature, rainfall and hours of sunshine were recorded each day in a town on the south coast of England.

These two **scatter graphs** were plotted from this data. Is it possible to work out the relationship between rainfall and hours of sunshine?

Key words

negative correlation

no correlation

positive correlation

scatter graphs

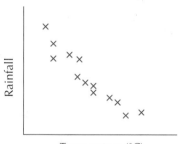

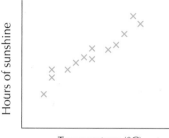

• The graph on the left shows **negative correlation**. In this case, it means that the higher the temperature, the less rainfall there is.

• The graph on the right shows **positive correlation**. In this case, it means that the higher the temperature, the more hours of sunshine there are.

Looking at both graphs together, what do they tell you about the effects of changes in temperature?

• The graph on the left indicates that high rainfall means low temperature.

• The graph on the right indicates that low temperature means little sunshine.

Putting this information together, you can deduce that high rainfall means little sunshine.

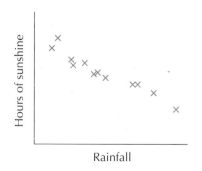

That is, rainfall and sunshine are negatively correlated, as shown in this graph.

Now look at this graph. It shows **no correlation** between the temperature and the number of fish caught daily off Rhyl – as you might expect.

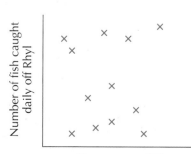

The table summarises the rules for combining two scatter graphs that have a common axis, to obtain the resulting correlation.

	Positive correlation	No correlation	Negative correlation
Positive correlation	Positive	No correlation	Negative
No correlation	No correlation	Cannot tell	No correlation
Negative correlation	Negative	No correlation	Positive

Notice, from the table, that the new graph can have its axes in either order, as this does not affect the correlation.

Exercise 4A

1 In a competition there are three different games.

a Ryan has played two games.

	Game A	Game B	Game C
Score	62	53	

To win, Ryan needs a mean score of 60.

How many points does he need to score in Game C?

b Ian and Nina have played all three games.

	Game A	Game B	Game C
Ian's scores	40		
Nina's scores	35	40	45

Their scores have the same mean.

The range of Ian's scores is twice the range of Nina's scores. Copy the table above and fill in the missing scores.

c These scatter diagrams are drawn from the scores of everyone who played all three games.

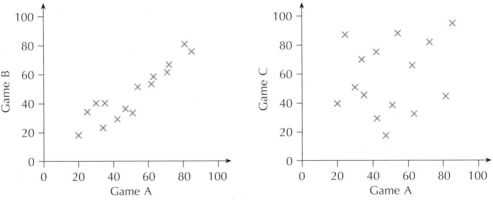

Look at the scatter diagrams and describe the relationship between:

i Game A and Game B   **ii** Game A and Game C   **iii** Game B and Game C.

2 A post office compares:

- the cost of postage with the mass of each parcel
- the cost of postage with the size of each parcel.

The results are shown on these scatter graphs.

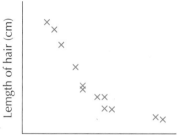

a Describe the type of correlation between the mass of parcels and the cost of postage.

b Describe the type of correlation between the size of parcels and the cost of postage.

c Describe the relationship between the mass of parcels and the size of parcels.

d Draw a scatter graph to show the correlation between the mass of parcels and the size of parcels. (Plot about ten points for your graph.)

3 A pupil compared the ages of a group of men with the length of their hair and also with their masses. His results are shown on the two scatter graphs.

a Describe the type of correlation between the length of hair and age.

b Describe the type of correlation between the length of hair and mass.

c Describe the relationship between age and mass.

d Draw a scatter graph to show the correlation between age and mass. (Plot about ten points for your graph.)

 4 Philip was looking at a magazine in his dentist's waiting room. He read that there were some correlations between various subjects often studied in schools.

When he got home he could only remember that there was:

- a positive correlation between mathematics and music
- a positive correlation between mathematics and physics
- a negative correlation between English and art
- no correlation between mathematics and art.

Explain how he could find the correlation between:

a music and physics   **b** English and mathematics   **c** art and physics.

PS **5** **a** The price of oil and the cost of petrol show a positive correlation. The cost of petrol and the price of food have a positive correlation.

What can you say about the price of oil and the price of food?

b The number of large coats sold has a negative correlation with the average daily temperature. The number of scarves sold has a negative correlation with the average daily temperature.

What can you say about the numbers of scarves and large coats sold?

c The number of newspapers sold has no correlation with the hours of sunshine. The number of cups of coffee sold in cafés has no correlation with the number of newpapers sold.

What can you say about the number of hours of sunshine and the number of cups of coffee sold in cafes?

6 Armand does a study and finds out that:

• the lower the temperature, the more spiders he sees in the house

• the fewer spiders he sees in the house, the happier his sister feels.

Explain any correlation between the temperature and how happy Armand's sister is.

Investigation: Comparing marks

Collect test marks for ten pupils in three different subjects: for example, mathematics, science and art.

Draw the scatter graphs for:

• mathematics and science

• mathematics and art

• science and art.

Comment on your results.

You could use a table like this to show the test marks.

	Pupil 1	Pupil 2	Pupil 3	Pupil 4	Pupil 5	Pupil 6	Pupil 7	Pupil 8	Pupil 9	Pupil 10
Maths										
Science										
Art										

4.2 Time-series graphs

Learning objective

• To use and interpret a variety of time-series graphs

Key word

time-series graph

A **time-series graph** is any graph that has a time scale.

Example 1

Look at the graphs below.

Try to match each graph to one of the statements listed after graph 5.

Graph 1: Mean temperature difference from normal for the UK

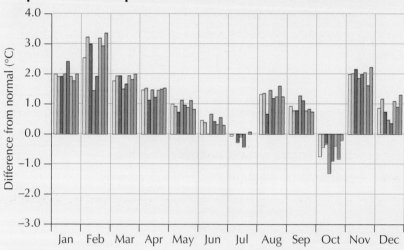

Graph 2: Average annual temperatures for central England, 1659–2010

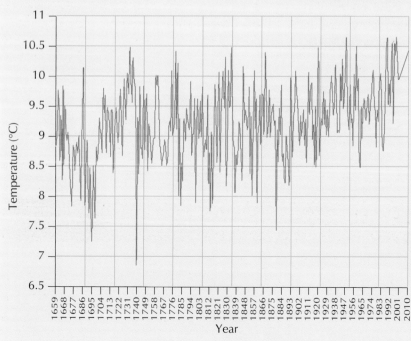

(Continued)

Graph 3: Monthly rainfall data for Perth, Australia

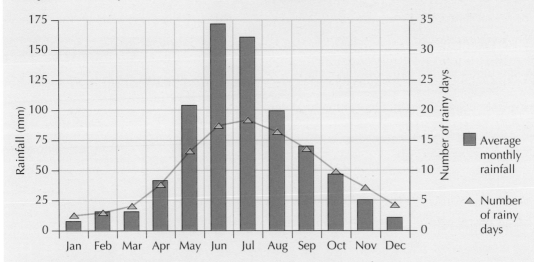

Graph 4: Monthly rainfall data for Brisbane, Australia

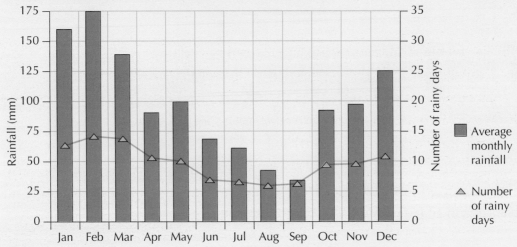

Graph 5: Monthly temperature data for Perth, Australia

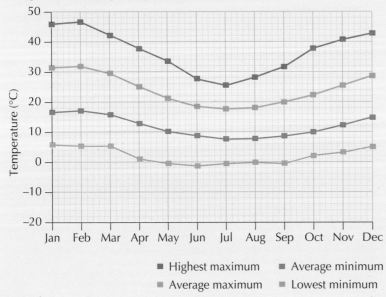

Statement A: February is the hottest month here.

Statement B: October was colder than normal.

Statement C: September is a fairly dry month.

Statement D: This country is gradually getting warmer.

Statement E: The difference between the lowest and highest temperatures is least in July.

Comparing the statements and the graphs, you can see that:

Statement A matches graph 5

Statement B matches graph 1

Statement C matches graph 4

Statement D matches graph 2

Statement E matches graph 5.

Exercise 4B

 1   The line on the graph below represents a speed of 40 mph.

    a   Copy the graph and draw on it a line to represent a speed of 20 mph.
Label the line by writing '20 mph'.

    b   Now draw a line on the graph to represent a speed of 70 mph.
Label the line by writing '70 mph'.

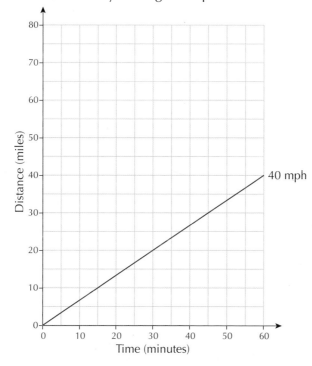

 The abbreviation 'mph' stands for 'miles per hour'.

(MR) **2** This time-series graph shows the height to which a ball bounces against the time after it was dropped.

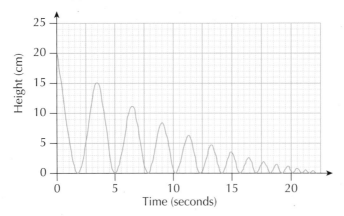

a Write a comment stating what happens to the length of time for which the ball is in the air after each bounce.

b This ball always bounces to a fixed fraction of its previous height.

What fraction is this?

c After how many bounces does this ball bounce to less than half of the greatest height?

d In theory, how many bounces does the ball make before it comes to rest?

(MR) **3** Look again at the graph showing the mean temperature changes in the UK.

Mean temperature difference from normal for the UK

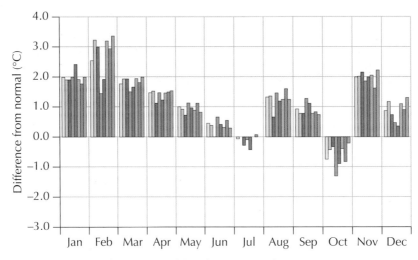

a Which months were colder than normal?

b Which month was much warmer than normal?

c In a short sentence, describe what the graph shows about the UK in that year.

4 Look again at the graphs showing the rainfall for Perth and Brisbane.

Monthly rainfall data for Perth, Australia

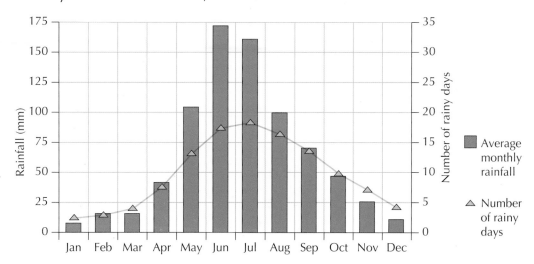

Monthly rainfall data for Brisbane, Australia

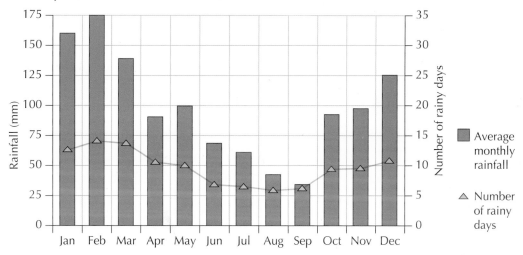

a Which month has the greatest rainfall in Perth?

b Which month has the least rainfall in Brisbane?

c Explain how you can tell that Perth and Brisbane are not in the same region.

d Which place, Perth or Brisbane, has more days of rainfall over the year?
Show how you worked it out.

5 The graph shows the change in numbers of licensed vehicles on the UK roads from 1995 to 2013.

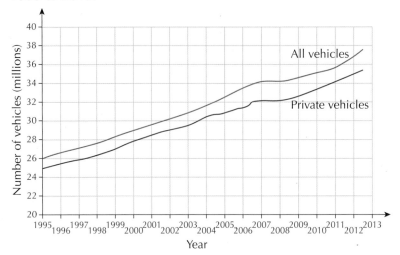

a In 1995, the department of transport were expecting the number of private licensed vehicles on the road to increase by over 40%.

Has this happened? Give reasons to support your conclusion.

b What percentage increase was there in the total number of licensed vehicles on the UK roads from:

i 2000 to 2006   **ii** 2006 to 2012?

c Use the graph to suggest how many private licensed vehicles there will be on the roads in the year 2014.

Investigation: Is England getting warmer?

Look again at the graph showing average annual temperatures for Central England from 1659 to 2001.

A Use the graph to write a report explaining how the England is getting warmer.

B Copy the latter part of the graph, from 1956 onwards, and extend it to 2028, showing what you expect to happen.

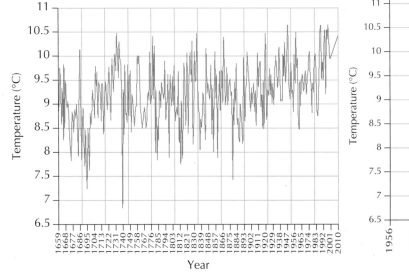
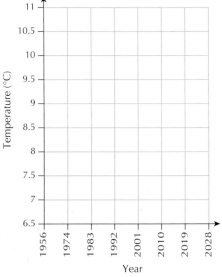

4.3 Two-way tables

Learning objective

- To interpret a variety of two-way tables

Key word

two-way table

As you study mathematics, you will see many different **two-way tables**, from timetables to data-analysis tables. This section will allow you to practise interpreting a variety of two-way tables.

Example 2

An internet company bases its delivery charges for its goods on the total cost of the order and on the type of delivery:

- normal delivery (taking 3 to 5 days)
- next-day delivery.

The table shows how the company calculates the charge.

		Delivery	
		Normal (3 to 5 days)	Next day
Cost of order (£)	0–10	1.95	4.95
	10.01–30	2.95	4.95
	30.01–50	3.95	6.95
	50.01–75	2.95	4.95
	Over 75	Free	3.00

a Comment on the difference in delivery charges for normal and next-day delivery.

b Two items cost £5 and £29 respectively. How much would you save by ordering them together:

   i using normal delivery   **ii** using next-day delivery?

     a It always costs more using next-day delivery but, for goods costing between £10.01 and £30 or between £50.01 and £75, it is only £2 more. It is £3 more for all other orders.

     b i Using normal delivery:

- ordering the items separately, it would cost £1.95 + £2.95 = £4.90
- ordering them together would cost £3.95.

       The saving would be £4.90 – £3.95 = 95p.

     ii Using next-day delivery:

- ordering the items separately, it would cost £4.95 + £4.95 = £9.90
- ordering them together would cost £6.95.

       The saving would be £9.90 – £6.95 = £2.95.

1 Look at this two-way table.

		Colour of car				
		Red	White	Blue	Black	Other
	Peugeot	8	1	4	1	4
	Ford	11	2	4	2	6
Make of car	Vauxhall	5	4	0	0	2
	Citroen	1	2	2	0	3
	Other	6	3	3	4	2

If a car is chosen at random, what is the probability that it is:

a a Peugeot   **b** red   **c** a red Peugeot

d not blue   **e** not a Ford   **f** a black Vauxhall?

2 The table shows the percentages of boys and girls, by age group, who have a mobile phone.

		Boys (%)	Girls (%)
	10	28	24
	11	31	28
	12	52	49
Age	13	63	66
	14	66	69
	15	72	74

a Comment on the percentage differences between boys and girls.

b Comment on any other trends that you notice.

3 The cost of a set of old toys depends on whether the toys are still in the original boxes and also on the condition of the toys. The table shows the percentage value of a toy compared with its value if it is in perfect condition and boxed.

		Boxed	Not boxed
	Excellent	100%	60%
	Very good	80%	50%
Condition	Good	60%	40%
	Average	40%	25%
	Poor	20%	10%

a Copy and complete the table.

		Difference between boxed and not boxed
	Excellent	100% − 60% = 40%
	Very good	
Condition	Good	
	Average	
	Poor	

b Explain the effect of the set being boxed compared with the condition of the toys.

(MR) **4** The table shows the heights of 70 Year 9 pupils. The measurements are given to the nearest centimetre.

Height (cm)	Boys	Girls
130–139	3	3
140–149	2	4
150–159	10	12
160–169	14	11
170–179	6	5

Use the results to examine the claim that, in Year 9, the boys are taller than the girls. You may use a frequency diagram to help you.

(PS) **5** The table shows the unemployment rates in some European countries, over a five-year period.

	Unemployment rates (%)				
	2009	2010	2011	2012	2013
France	9.2	9.4	9.3	10.0	10.5
Germany	7.8	7.1	5.9	5.5	5.5
Italy	7.9	8.5	8.5	10.8	12.0
Netherlands	3.7	4.5	4.5	5.3	6.2
Sweden	8.3	8.5	7.7	7.9	8.1
United Kingdom	7.6	7.9	8.1	8.0	7.9

a Create a time-series graph to compare the rates of unemployment in these six countries.

b Write a report on what you notice.

Problem solving: Birth months

A school analysed the information about the birth months of 1000 pupils.

The results are shown in the table.

Month	Jan	Feb	Mar	Apr	May	Jun	Jul	Aug	Sep	Oct	Nov	Dec
Boys	34	36	43	39	47	50	44	39	55	53	42	35
Girls	37	31	36	35	44	43	36	40	52	49	43	37

A On the same grid, plot both sets of values to give a time-series graph for the boys and another for the girls.

B Use the graphs to examine the claim that more children are born in the summer than in the winter.

4.4 Comparing two or more sets of data

Learning objective

- To compare two sets of data from statistical tables and diagrams

Two cars, A and B, each cost £20 000 when they were new.

The graphs show how the values of the cars fell over the next eight years.

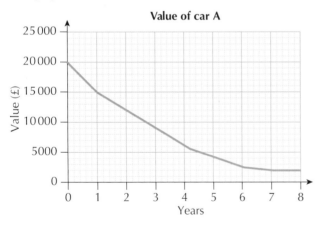

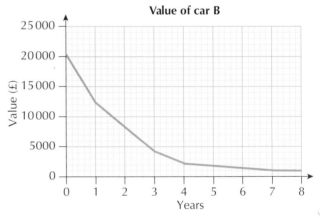

Which car's value fell more?

How can you tell?

Example 3

A teacher is comparing the reasons for the absence of pupils who have had time off school.

The charts show the reasons for absence of two different year groups.

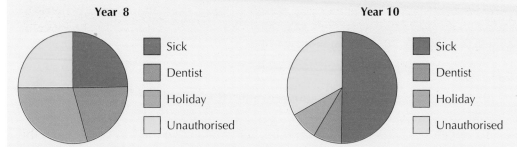

Year 8

Year 10

- Sick
- Dentist
- Holiday
- Unauthorised

One hundred pupils in Year 8 and 40 pupils in Year 10 had time off school.

The teacher says: 'The charts show that more pupils in Year 10 than in Year 8 were absent because they were sick.'

Explain why the charts do not show this.

The number of Year 8 pupils absent because they were sick was a quarter of 100, which is 25.

The number of Year 10 pupils absent because they were sick was half of 40, which is 20.

So fewer Year 10 pupils than Year 8 pupils were absent because they were sick.

Exercise 4D

 1 The graph shows the attendance at two concerts, a classical concert and a rock concert.

Comment on the proportions of children attending each concert.

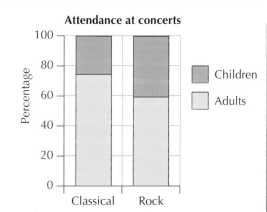

Attendance at concerts

- Children
- Adults

 2 One hundred pupils took two tests: a science test and a mathematics test. The results are shown on the graph.

Which test did the pupils find more difficult? Explain your answer.

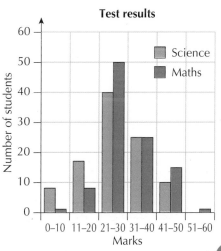

Test results

- Science
- Maths

3 The chart shows the percentages of trains that were on time and late during one day.

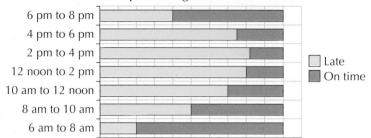

☐	Late
■	On time

a Compare the lateness for different parts of the day.

b Comment on what you would expect to happen between 8 pm and 10 pm.

 4 Here are two sets of masses of a sample of two different makes of 400 g chocolate bars.

Chunky Bar (g)	Choctastic (g)
401	391
407	410
405	407
404	402
403	413
404	395

a Calculate the mean and range of the data for each sample.

b Draw charts to compare the two makes of chocolate.

c Comment on which one you would buy.

5 The table shows the circulation figures for the major UK national newspapers over a twenty-year period.

	1993	2003	2013
The Sun	3 690 000	3 578 000	2 410 000
The Mirror	2 910 000	2 071 000	1 058 000
The Daily Mail	1 700 000	2 519 000	1 863 000
The Daily Express	1 500 000	983 000	530 000
The Daily Telegraph	1 060 000	947 000	556 000
The Daily Star	859 000	835 000	536 000
The Times	390 000	671 000	399 000
The Guardian	412 000	410 000	204 000
The Independent	376 000	221 000	76 802

a Comment on the trend of buying daily newspapers over the last twenty years. Why do you think this is?

b Which newspaper had the greatest percentage loss of readers?

c Draw a diagram illustrating this changing pattern of buying newspapers.

Activity: Howzatt!

The chart shows the number of times the different counties have won the County Cricket Championship from 1864 to 2013.

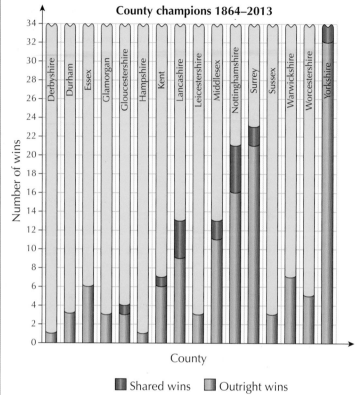

County champions 1864–2013

A Find out where these counties are on a map of Britain and comment on which part of the country appears to have been dominant, in terms of cricket.

B Are there any exceptions?

4.5 Statistical investigations

Learning objective

• To plan a statistical investigation

Investigating a problem will involve several steps.

Look at the three examples described in this section. They are taken, from different subjects, following an overall plan.

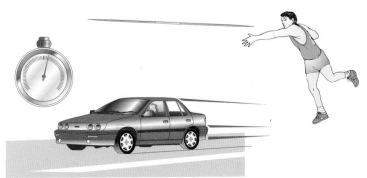

	Step	PE	Science	Geography
1	Decide which general topic to study.	How to improve pupil performance in sport	Effect of engine size on a car's acceleration	Life expectancy versus cost of housing
2	Specify in more detail.	A throwing event	A particular make of car	Compare house prices in Yorkshire with those in south-east England.
3	Consider questions that you could investigate.	How much further do pupils throw using a run up? Is a Year 9 pupil able to throw as far as a Year 11 pupil of the same height?	Does a bigger engine always mean that a car can accelerate faster?	Do people who live in expensive housing tend to live longer?
4	State your hypotheses (your guesses at what could happen).	Distance thrown will improve if using a run-up. Year 11 pupils of the same height may be physically stronger and would therefore throw further.	In general, more powerful engines produce the greater acceleration. More powerful engines tend to be in heavier cars and therefore the acceleration will not be affected. Larger engines in the same model of car will improve acceleration.	People in expensive housing have greater incomes and also may have a longer life expectancy.
5	State your sources of information required.	Survey of distance thrown with different lengths of run-up	Magazines and/or books with information on engine sizes and acceleration times for 0–60 mph	Library or the internet for census data for each area
6	Describe the relevant data.	Choose pupils from different age groups with a range of heights. Make sure that there are equal numbers of boys and girls. Choose pupils from the full range of abilities.	Specify make of car, engine size and acceleration. Note: The government requires car manufacturers to publish the time taken to accelerate from 0 to 60 mph.	Average cost of housing for each area Data about life expectancy for each area
7	List possible problems.	Avoid bias when choosing your sample or carrying out your survey.	Petrol engines must be compared with other petrol engines, not with diesel engines.	

8 Identify method of data collection.	Make sure that you can record all the factors that may affect the distance thrown: for example, age or height.	Make sure that you can record all the information that you need, such as engine size and weight of car. Remember to quote sources of data.	Extract relevant data from sources. Remember to quote sources of data.
9 Decide on the level of accuracy required.	Decide how accurate your data needs to be; for example, nearest 10 cm.	Round any published engine sizes to the nearest 100 cm^3 (usually given as 'cc' in the car trade), which is 0.1 litre; for example, a 1905 cc engine has a capacity of approximately 1.9 litres.	
10 Determine sample size.	Remember that collecting too much information may slow down the experiment.		
11 Construct tables for large sets of raw data, in order to make work manageable.	Group distances thrown into intervals of 5 metres. Use two-way tables to highlight differences between boys' and girls' data.		Group population data in age groups of, for example, 10-year intervals.
12 Decide which statistics are most suitable.	When the distances thrown are close together, use the mean. When there are a few extreme values, use the median.		Sample should be sufficiently large to be able to use the mean.

Exercise 4E

  1 Look at the three examples presented above.

Investigate *either* one of the examples given *or* a problem of your own choice. You should follow the steps given in the table, including your own ideas.

Investigation: A comparison

Think of a problem related to a piece of work, such as a foreign language essay or a history project. See whether you can use the step-by-step plan to carry out a statistical investigation.

For example, you may wish to compare the word lengths of an English and a French piece of writing, or you may wish to compare data about two wars.

Ready to progress?

I know how to interpret graphs and charts, and how to draw conclusions.

I know how to draw conclusions from scatter graphs and I have a basic understanding of correlation.
I know how to interpret a variety of two-way tables.
I know how to compare two sets of data and draw conclusions.

I know how to generate a detailed solution to a problem.
I know how to compare distributions and comment on what I find.
I know how to use measures of average and range to compare distributions and make inferences.

Review questions

 1 A local newspaper story about public libraries in its region included this diagram.

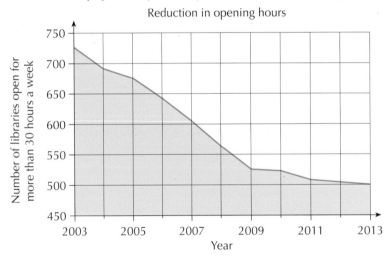

Reduction in opening hours

Use the diagram to decide whether each of these statements is true or false, or whether you are uncertain.

a The number of libraries open for more than 30 hours per week fell by more than half from 2003 to 2013.

Explain your answer.

b In 2017 there will be about 450 libraries open for more than 30 hours a week in this region.

Explain your answer.

2 The graph below gives information about the diameters and heights of a sample of three types of tropical fruit. Use the dotted lines on the graph to decide which type of fruit each point represents.

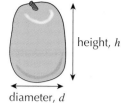

a The diameter of a fruit of type C is 12 cm.

What would you expect its height to be?

b The diameter of a different fruit is 3.8 cm. Its height is 8.5 cm.

Which of the three types of fruit is it most likely to be?

Explain your answer.

c Which type of fruit is most nearly spherical in shape?

Explain your answer.

d You can find the approximate volume of a fruit by using the formula:

$$V = \tfrac{1}{6}\pi d^2 h$$

where V is the volume, d is the diameter and h is the height.

The diameter of a fruit is 4.3 cm and its height is 5.4 cm.

What is the approximate volume of this fruit?

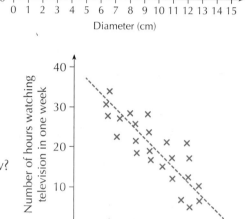

3 A teacher investigated whether pupils who watch less television study more.

The scatter graph shows his results. He drew a line on the graph to illustrate the trend.

a What type of correlation does the graph show?

b One of the teacher's pupils said the equation of the line drawn is $y = x + 35$.

Explain how you can tell that this equation is wrong.

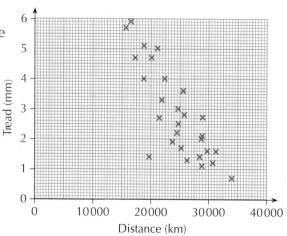

4 Car tyres are unsafe if their tread is too worn away.

The tread depth on a tyre and the distance travelled by that tyre were measured for a sample of tyres. The results are shown on this scatter graph.

Tyres with a tread depth of less than 1.6 mm are illegal.

The government proposes to change this to a depth of less than 2.5 mm.

a How would you expect this to affect the distance a tyre would last before it became illegal? Give your answer in kilometres.

b How many of these tyres would now be illegal?

Challenge

Rainforest deforestation

Since 1970, over 600 000 km^2 of Amazon rainforest have been destroyed. This is an area larger than Spain.
Between the years 2000 and 2005, Brazil lost over 132 000 km^2 of forest – an area about the same size as England.
Between the years 2005 and 2013, Brazil lost over 90 000 km^2 of forest – an area about the same size as Scotland.
The table below shows how much of the rainforests in Brazil have been lost each year since 1988.

Deforestation figure	
Year	**Deforestation (sq km)**
1988	21 000
1989	18 000
1990	14 000
1991	11 000
1992	14 000
1993	15 000
1994	15 000
1995	29 000
1996	18 000
1997	13 000
1998	17 000
1999	17 000
2000	18 000
2001	18 000
2002	21 000
2003	25 000
2004	27 000
2005	19 000
2006	14 000
2007	10 000
2008	9 000
2009	7 000
2010	6 000
2011	5 000
2012	6 000

Use the information on the opposite page to answer these questions.

1 From 1988 to 1991, Brazil had an economic slowdown. What was happening to the rate of deforestation during that time?

2 From 1992 to 1995, Brazil had economic growth. What was happening to the rate of deforestation during that time?

3 What do think was happening to Brazil's economy:

   a from 1998 to 2004   **b** from 2005 to 2012?

4 What does the chart and the information given in questions **1** and **2** suggest about the link between deforestation in Brazil and the economy?

5 Draw a graph, showing the total deforestation since 1988, year by year.

6 If the rate of deforestation over 2005–12 continued at the same rate, estimate when the deforestation would be 1000 km^2.

The pie chart below shows the three main reasons for deforestation in the Amazon during the period 2000–13.

7 What percentage of the deforestation was caused by each reason?

8 It was suggested that over the next three years:

   🌳 the same amount of deforestation would take place

   🌳 the amount of construction work would actually double

   🌳 the number of small farms would halve

   🌳 the number of cattle ranches would increase.

Draw a pie chart reflecting the reasons for deforestation suggested for 2016.

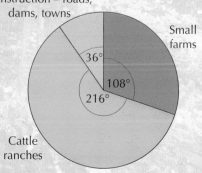

Deforestation in the Amazon, 2000–2013

Construction – roads, dams, towns

Small farms

36°

108°

216°

Cattle ranches

5

Applications of graphs

This chapter is going to show you:

- how to interpret and draw step graphs
- how to interpret and draw time graphs
- how to interpret and draw exponential growth graphs.

You should already know:

- how to read information from graphs
- how to draw graphs on coordinate axes.

About this chapter

Real-life graphs are used to improve the performance of racing-car drivers. Their speed, at various points on the racing track, is plotted on a graph, which is then analysed by the driver's team.

Sophisticated ways to use computers and electronic devices in the car are most evident in the design and construction of Formula 1 racing cars. These include:

- sensors for measuring the pressure and temperature of tyres, engine temperature and oil pressure
- regulators for dispensing of fuel
- highly automated communication systems by which measurements are made for analysis of the car through graphs and charts.

5.1 Step graphs

Learning objective

• To interpret step graphs

Key words

discontinuous graph

step graph

A **step graph** is a special type of line graph that is made up of horizontal lines in several intervals or steps.

You would generally use a step graph to represent situations that involve sudden jumps across intervals, such as cost of postage, car parking fees or telephone rates.

Example 1

This step graph shows a car park's charges.

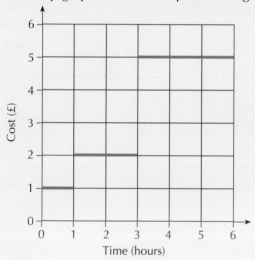

a What is the charge for parking a car for:

   i   30 minutes   **ii** less than 1 hour   **iii** 2 hours

   iv 2 hours 59 minutes   **v** 3 hours 30 minutes   **vi** 6 hours?

b How long can I park for:

   i £1   **ii** £2   **iii** £5?

You can read the answers from the graph.

  a i £1   **ii** £1   **iii** £2   **iv** £2

    v £5   **vi** £5

  b i Up to 1 hour   **ii** Up to 3 hours

    iii Up to 6 hours

Hint The value at the end of each step is usually included.

Use graph paper to draw the graphs in this exercise.

1 This step graph shows how the cost of a single ticket on a transport system varies.

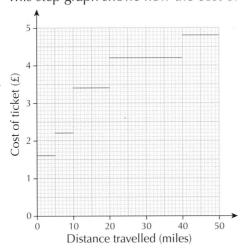

What is the price of a ticket to travel:

a 3 miles   **b** 10 miles   **c** 32 miles   **d** 50 miles?

 2 House owners have to pay council tax every year.

Each property is put into a band, depending on its value in 1991.

The step graph shows, to the nearest £100, how much council tax was paid, per year, to a city council for the year 2013–2014 for each band.

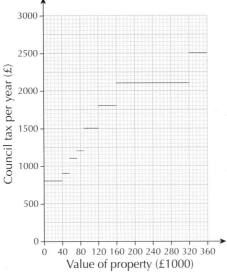

a How much council tax was paid per year for a property with a value of:

     i £80 000   **ii** £116 000   **iii** £299 50   **iv** £355 500?

b Council tax is paid in 10 monthly instalments.

A property has a value of £150 000 and the actual council tax is £1801.46 for the year.

The first instalment is £181.46 and the rest of the instalments are equally divided over the remaining months.

Work out the amount of council tax for each of the remaining months.

3 The table shows a country's parcel post costs.

Mass, m (kg)	Home country	Abroad
$0 < m \leqslant 0.5$	£1.40	£3.50
$0.5 < m \leqslant 1$	£2.50	£4.60
$1 < m \leqslant 2$	£3.20	£5.90
$2 < m \leqslant 3$	£4.60	£7.00
$3 < m \leqslant 5$	£5.50	£9.50
$5 < m \leqslant 10$	£6.00	£12.00
$10 < m \leqslant 20$	£8.00	£15.00

Hint Remember that
0.5 kg = 500 grams

Draw step graphs to show charges against mass for:

a the home country **b** abroad.

(FS) 4 A taxi's meter reads £2 at the start of every journey. As soon as the taxi has travelled two miles, £3 is added to the fare. The reading on the meter then increases in steps of £3 for each whole mile covered, up to five miles. For journeys over five miles, an extra £1 is added per mile.

a What fare will be charged for a journey of:

   i half a mile   **ii** 1 mile   **iii** 3 miles

   iv 5 miles   **v** 6 miles   **vi** 10 miles?

b Draw a step graph to show the fares for journeys up to 10 miles.

5 The table shows the costs of sending a large letter by first-class and second-class post.

Mass of letter (g)	First-class post (£)	Second-class post (£)
0–100	0.90	0.69
101–250	1.20	1.10
251–500	1.60	1.40
501–750	2.30	1.90

Draw step graphs to show cost against mass for:

a first-class post **b** second-class post.

(MR) 6 Match the four graphs shown here to the situations described below.

a The amount John gets paid against the number of hours he works

b The temperature of an oven against the time it is switched on

c The amount of tea in a cup as it is drunk

d The cost of posting a letter compared to the mass

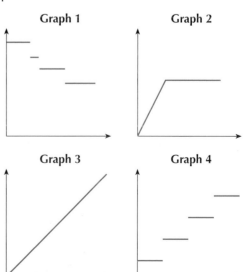

Graph 1   Graph 2

Graph 3   Graph 4

Reasoning: Discontinuous graphs

Step graphs are sometimes called **discontinuous graphs.**

They are often given in this format.

Each horizontal step has an open circle ○ at the start and a coloured circle ● at the end point.

The value marked by an open circle is excluded from that step.

The value marked by a coloured circle is included in that step.

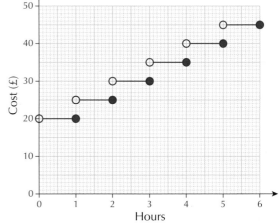

Athletic event starting times
(by age bracket)

A This discontinuous graph shows the cost of hiring a van for up to six hours.

How much does it cost to hire the van for:

a $1\frac{1}{2}$ hours   **b** 3 hours 20 minutes

c $4\frac{1}{4}$ hours   **d** 2 hours

e 5 hours   **f** 50 minutes?

B This discontinuous graph shows how the parking charges in a town's car park vary.

a How much will it cost to park for:

   i 30 minutes   **ii** $1\frac{1}{2}$ hours

   iii 3 hours   **iv** 4 hours 40 minutes?

b How much will it cost to park for 7 hours?

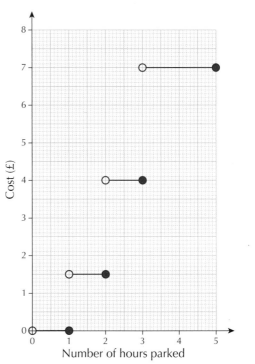

5.2 Time graphs

Learning objective

- To interpret and draw time graphs

Key words

distance–time graph

time graph

Time graphs are used to show a relationship that exists between a variable and time.

You will see time graphs used to show:

- how a distance changes with time, usually referred to as **distance–time graphs**
- how temperature changes with time
- how populations change with time.

Example 2

The time graph shows how the temperature of water increases with time, as it is heated.

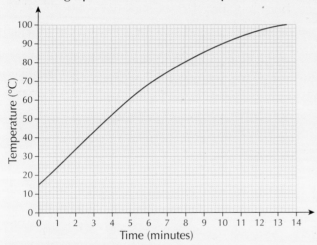

a Estimate the temperature of the water after:

  i $4\frac{1}{2}$ minutes   ii $9\frac{1}{2}$ minutes.

b Estimate how long it takes for the water to boil (when the temperature reaches 100 °C).

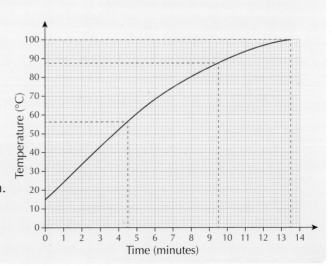

  a See the red lines drawn on the graph.

   i After $4\frac{1}{2}$ minutes the temperature is about 56 °C.

   ii After $9\frac{1}{2}$ minutes the temperature is about 88 °C.

  b See the green lines drawn on the graph.

   It takes $13\frac{1}{2}$ minutes for the water to boil.

Example 3

The distance–time graph below illustrates three people running in a race.

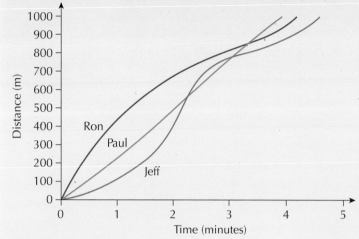

The graph shows how quickly each person ran, who was ahead at various times, who won and by how many seconds. Comment on each graph.

Paul: Notice that Paul's graph is a straight line. This means that he ran at the same speed throughout the race. Paul won the race, finishing about 20 seconds before Ron.

Ron: The shape of Ron's graph indicates that he started quickly and then slowed down. He was in the lead for the first 850 metres, before Paul overtook him.

Jeff: Jeff started slowly, but then picked up speed to overtake Paul, staying ahead of him for a minute before running out of steam and slowing down to come in last, about 30 seconds behind Ron.

 The steeper the graph, the faster the person is running.

Exercise 5B

1 This distance–time graph shows how two rockets flew during a test flight. Rocket D flew higher than Rocket E.

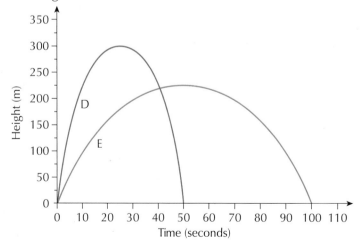

a Estimate the height reached by Rocket D.

b Estimate how much higher Rocket D went than Rocket E reached.

c How long after the launch were both rockets at the same height?

d For how long was each rocket higher than 150 m?

e Can you tell which rocket travelled further? Explain your answer.

2 This graph illustrates the amount of water in a bath after it has started to be filled.

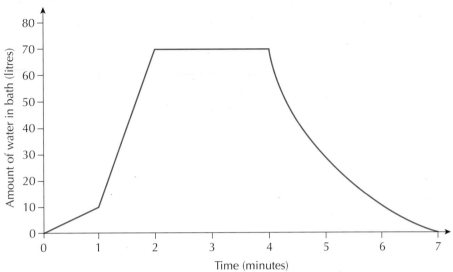

a Explain what might have happened one minute after the start.

b When was the plug pulled out, for the bath to start emptying?

c Why do you think the graph shows a curved line while the bath was emptying?

d How long did the bath take to empty?

3 This distance–time graph shows the journey of a jogger on a 5-mile run. At one point she stopped to admire the view and at another point she ran up a steep hill.

 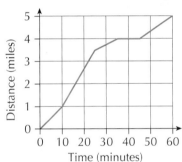

a For how long did she stop to admire the view?

b What distance into her run was the start of the hill?

4 The graph shows how a liquid cools down.

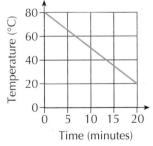

a How long does it take the liquid to cool down to 20 °C?

b Estimate the temperature of the liquid after 10 minutes.

c Estimate how long it takes the liquid to cool from 60 °C to 40 °C.

5 Ari goes on a cycle ride to visit two of his friends.

The time–distance graph shows his journey.

  a At what time did he arrive at the first friend's house?

  b How long did he stay at the first friend's house?

  c At what time did he arrive at the second friend's house?

  d How long did he stay at the second friend's house?

  e How many kilometres did he cycle altogether?

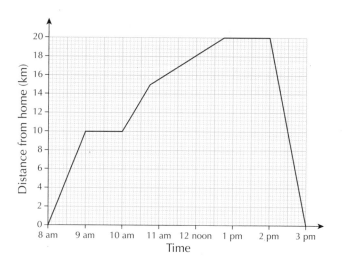

6 Match the three distance–time sketch graphs shown here to the situations described below.

Graph 1

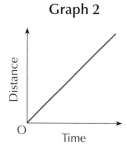

Graph 2

Graph 3

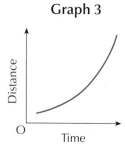

  a The distance travelled by a train moving at a constant speed

  b The distance travelled by a motorbike accelerating to overtake

  c The distance travelled by an old car, which starts well, but gradually slows down

PS **7** Water drips steadily into the container shown. The sketch graph shows how the depth of water varies with time.

 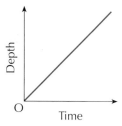

Sketch similar sketch graphs for bottles with these shapes.

  a   **b**   **c**   **d**

Problem solving: Population increase in the UK

The UK population has been increasing over the last 200 years.

This table shows the population every 20 years.

A Draw a graph to show how the population has increased since 1801.

B From your graph, estimate what the population was in 1951.

C Calculate the percentage population increase from:

    a 1881 to 1901   **b** 1981 to 2001.

    Give your answers to one decimal place.

D Can you predict the UK population for 2021? Give a reason for your answer.

Year	Population (millions)
1801	12
1821	15.5
1841	20
1861	24.5
1881	31
1901	38
1921	44
1941	47
1961	53
1981	56
2001	59

5.3 Exponential growth graphs

Learning objective

- To interpret and draw exponential growth graphs

Key word

exponential growth graph

Graphs that continuously increase at a fixed rate are known as **exponential growth graphs**.

Graphs that show an increase in population or the increase in investments at a bank are examples of exponential growth graphs.

Example 4

The population of a village doubles every five years.

The table shows the increase in the population over 20 years.

Number of years	0	5	10	15	20
Population	500	1000	2000	4000	8000

Estimate the number of years it will take for the population of the village to reach 5000.

Plot the points on a graph.

Draw a smooth curve passing through all the points.

Then draw suitable lines on the graph, as shown, to estimate the number of years for the population of the village to reach 5000.

Start with a horizontal line from 5000, then draw a vertical line down, from where the first line meets the curve.

This is $16\frac{1}{2}$ years.

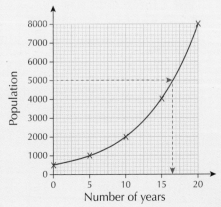

1 The population growth over time (measured in seconds) of a particular bacteria is shown on the graph.

    a How many bacteria are alive after one minute?

    b How many bacteria are alive after two minutes?

    c How many bacteria are alive after three minutes?

    d How long did it take the population of bacteria to reach 30 000?

    e How long did it take the population of bacteria to reach 70 000?

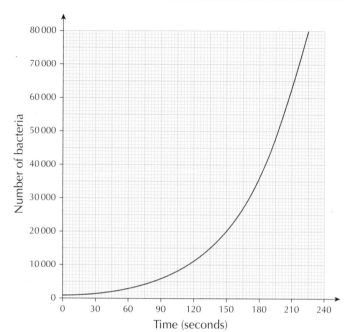

(FS) **2** The graph shows how an investment has grown exponentially over a 30-year period.

    a What was the initial investment?

    b How long did it take for the investment value to grow to £20 000?

    c How long did it take for the investment value to grow to £50 000?

    d What was the value of the investment after 20 years?

    e What was the value of the investment after 30 years?

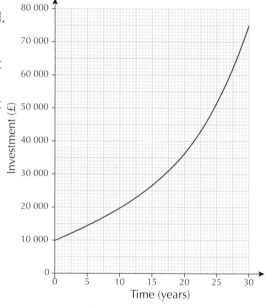

3 The graph shows how fast a truck is travelling as it increases its speed.

    a What is the truck's maximum speed?

    b What is the speed of the truck after 11 seconds?

    c How many seconds does it take the truck to reach a speed of 50 km/h?

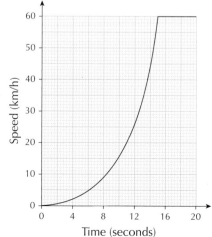

4 The graph shows the average growth rate of black sea turtles.

    a Estimate the length of a black sea turtle when it is born.

    b Estimate the length of a black sea turtle when it is 40 years old.

    c Estimate the age of a black sea turtle when it has grown to 40 cm.

    d Estimate the age of a black sea turtle when it has grown to 80 cm.

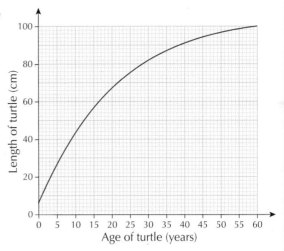

5 A population of rabbits is introduced into a new habitat. The table shows the growth of the population over a five-year period.

Year	0	1	2	3	4	5
Number of rabbits	100	150	225	338	506	759

    a Draw a graph to show the increase in the rabbit population.

    b How many rabbits were originally introduced?

    c Describe how the rabbit population is increasing.

    d If the rabbit population growth continues at the same rate, estimate how many rabbits there will be after six years.

Reasoning: Exponential decay

Each year a school holds a tennis tournament. The tournament starts with 128 players. During each round, half of the players are eliminated.

A Copy and complete the table to show how many players are left after each round.

Round	1	2	3	4	5	6
Players left	64					

B How many players remain after six rounds?

C Plot the points on a graph to show the number of players left after each round.

D Explain why you cannot join the points with a curve.

Ready to progress?

I can interpret and draw step graphs.
I can interpret and draw time graphs.

I can interpret more complex time graphs.
I can interpret and draw exponential growth graphs.

Review questions

1 This step graph shows the cost of sending a letter from the UK to Europe.

Copy and complete the table below. The first row has been done for you.

Mass of letter (g)	Cost (p)
0–20	88

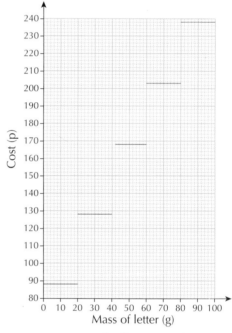

2 Rhiann and Ben ran a 1500 metre race. The distance–time graph illustrates the race.

Look carefully at the graph and then fill in the missing information in the report about the race.

Just after the start of the race, Rhiann was in the lead.

At 600 metres, Rhiann and Ben were level.

Then Ben was in the lead for ... minutes.

At ... metres, Rhiann and Ben were level again.

... won the race. The time was ... minutes.

... finished ... minute later.

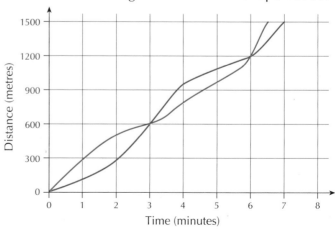

3   Harry decides to take a bath.

The graph shows the depth of water in the bath.

Match each section of the graph to the events below.

 i   Harry tops up the bath with hot water.
 ii   Harry lays back for a soak.
 iii Harry gets in the bath.
 iv   The hot and cold taps are turned on.
 v   The cold tap is turned off and only the hot tap is left on.
 vi   Harry washes himself.
 vii Harry pulls the plug and the bath empties.
 viii Harry gets out of the bath
 ix   The hot tap is turned off.

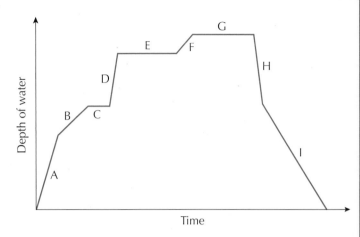

4   The time graph shows the speed of a car over a 50-second time period.

 a   Write down the fastest speed of the car.
 b   Explain what is happening between A and B.
 c   Explain what is happening between B and C.
 d   Explain what is happening between C and D.
 e   Explain what is happening between D and E.
 f   Explain what is happening between E and F.

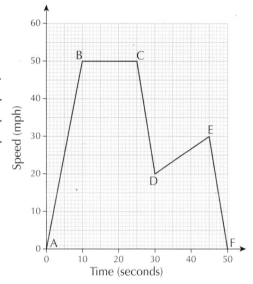

5   The population of a town increases by 10% every 10 years.

 a   Copy and complete the table to show the population after 40 years.

Number of years	0	10 years	20 years	30 years	40 years
Population	100 000				

 b   Draw an exponential growth graph to show this data.
 c   Use the graph to estimate the population of the town after 25 years.

6   a   Copy and complete the table for the graph of $y = 2^x$ for values of x from 0 to 6.

x	0	1	2	3	4	5	6
y	1			8			64

 b   On graph paper, draw the graph of $y = 2^x$ for values of x from 0 to 6.
 c   What type of graph have you drawn?

Problem solving
Mobile phone tariffs

Pick your plan						
Mix & match plans		Inclusive UK minutes and texts	Free 3 to 3 minutes	Free voicemail	Free Instant Messaging	Free Skype
£12 a month	£12 Promotional Tariff	100 Anytime any network minutes or texts or any mix of the two	300 3 to 3 minutes	✓	✓	✓
£15 a month	Mix & match 300	300 Anytime any network minutes or texts or any mix of the two	300 3 to 3 minutes	✓	✓	✓
£18 a month	Mix & match 500	500 Anytime any network minutes or texts or any mix of the two	300 3 to 3 minutes	✓	✓	✓
£21 a month	Mix & match 700	700 Anytime any network minutes or texts or any mix of the two	300 3 to 3 minutes	✓	✓	✓
£24 a month	Mix & match 900	900 Anytime any network minutes or texts or any mix of the two	300 3 to 3 minutes	✓	✓	✓
£27 a month	Mix & match 1100	1100 Anytime any network minutes or texts or any mix of the two	300 3 to 3 minutes	✓	✓	✓

- Pay by voucher or direct debit
- If you pay by voucher, service is suspended once the monthly allowance is reached
- 10% discount if you pay by direct debit
- If you pay by direct debit, any minutes or texts over the allowance are charged at 15p per minute or 15p per text
- All tariffs exclude VAT, which is charged at 20%

Use the information about Mobile phone tariffs to answer these questions.

1 How many hours is 900 minutes?

2 How many hours and minutes is 700 minutes?

3 Freya has the '£12 promotional' tariff. She pays by voucher. She does not use her phone for voice calls. So far in a month she has sent 43 texts. How many more can she send that month before service is suspended?

4 Tim has the '£12 promotional' tariff. He pays by direct debit. How much will this cost per month (excluding VAT) after the 10% discount?

5 Ben has the 'Mix and match 300' tariff. Before any discounts or VAT how much would this cost for a 12-month contract?

6 Hamza has the '£12 promotional' tariff. So far he has used 24 minutes on voice calls and sent 19 texts. How many more minutes or texts can he send before he gets charged extra?

7 Jo has the 'Mix and match 700' tariff and pays by voucher each month.

   a How much does this cost a month before VAT is added?

   b How much does this cost a month when VAT at 20% is added?

8 Zara has the 'Mix and match 500' tariff. She uses all of her any network minutes and all of her free '3 to 3' minutes. How many minutes did she talk in total? Give your answer in hours and minutes.

9 Mia has the 'Mix and match 500' tariff. She pays by direct debit. In one month she uses all of her allowed anytime, any network voice minutes and makes a further x minutes of voice calls. Her bill for the month before discount and VAT is £24. What is the value of x?

10 Tom has the 'Mix and match 700' tariff. He pays by direct debit. In one month he makes 550 anytime, any network voice minutes and sends y texts. His bill for the month before discount and VAT is £30. What is the value of y?

6

Pythagoras' theorem

This chapter is going to show you:

- how to use Pythagoras' theorem to calculate the lengths of sides in right-angled triangles
- how to use Pythagoras' theorem to solve problems.

You should already know:

- how to square a number
- how to calculate the square root of a number.

About this chapter

Thousands of years ago, the builders of the pyramids in Egypt used a rope with 12 equally spaced knots, which, when stretched out as shown here, formed a right-angled triangle.

This helped the builders make sure that any right angles required in the construction of the Pyramids were accurate.

The Great Pyramid of Giza in Egypt was finished more than 4500 years ago. It is the only one of the Seven Wonders of the World that still remains today. The pyramid was the largest of the Great Pyramids. Archaeologists believe that 2.3 million limestone blocks, around 2495 kilograms each, were put in place by from 20 000 to 100 000 labourers working to earn tax money after finishing the work of the harvest. Even with that many people, however, some ingenuity was required in the absence of today's machines and mechanics. The blocks, set without mortar, were fitted so tightly that there was no room even for a knife blade.

6.1 Introducing Pythagoras' theorem

Learning objective

• To understand Pythagoras' theorem

Key words

hypotenuse	Pythagoras
Pythagoras' theorem	

Pythagoras was a Greek philosopher and mathematician, who was born in about 581BC, on the island of Samos, just off the coast of Turkey. A very famous theorem about right-angled triangles is attributed to him.

In a right-angled triangle, the longest side, opposite the right angle, is called the **hypotenuse**.

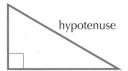

Exercise 6A

1 **a** Draw this right-angled triangle accurately, in the middle of a sheet of plain paper.

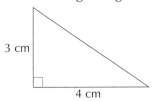

    b Measure the length of the hypotenuse. You should find it is exactly 5 cm.

    c Now draw squares on the three sides of the triangle and work out their areas.

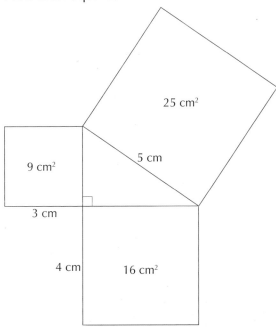

    d You should notice that the area of the square on the hypotenuse is equal to the sum of the areas of the other two squares.

$$5^2 = 3^2 + 4^2$$

2 **a** Make accurate drawings of these three right-angled triangles.

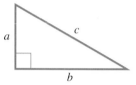

b Next, measure the length of the hypotenuse of each one.

Then copy and complete this table.

a	b	c	a^2	b^2	c^2
3	4				
5	12				
6	8				

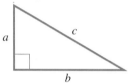

c Can you see a pattern in the last three columns? If you can, you have just rediscovered **Pythagoras' theorem**.

> In any right-angled triangle, the square of the hypotenuse is equal to the sum of the squares of the other two sides.
>
> Pythagoras' theorem is usually written as:
> $$c^2 = a^2 + b^2$$

3 **a** Draw this right-angled triangle accurately.

b Measure the length of the hypotenuse.
You should find it is 6.4 cm.

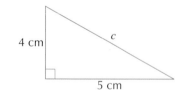

c Using Pythagoras' theorem, $c^2 = a^2 + b^2$:
$$c^2 = 4^2 + 5^2 = 16 + 25 = 41$$
So $c^2 = 41$.

To calculate c, you need to work out $\sqrt{41} = 6.4$ cm to one decimal place (1 dp).

So Pythagoras' theorem works.

6.2 Calculating the length of the hypotenuse

Learning objective

• To calculate the length of the hypotenuse in a right-angled triangle

Carefully follow through the next two examples. They will show you how to calculate the length of the hypotenuse in a right-angled triangle.

Example 1

Calculate the value of x in this triangle.

Using Pythagoras' theorem:

$$x^2 = 6^2 + 5^2$$
$$= 36 + 25$$
$$= 61$$

So $x = \sqrt{61} = 7.8$ cm (1 dp).

You should be able to work this out on a scientific calculator.

Try this sequence of keystrokes.

This may not work on every calculator. You may need to ask your teacher to help you.

Example 2

Calculate the length of the side labelled x in this triangle.

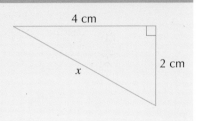

Using Pythagoras' theorem:

$$x^2 = 2^2 + 4^2$$
$$= 4 + 16 = 20$$

So $x = \sqrt{20} = 4.5$ cm (1 dp).

Exercise 6B 🖩

1   Calculate the length of the hypotenuse in each right-angled triangle.
Give your answers correct to one decimal place.

a   **b**   **c**

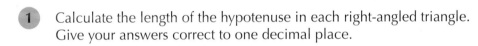

d   **e**   **f**

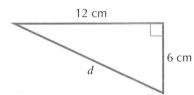
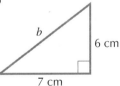
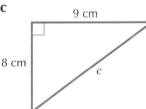

g   **h**

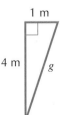
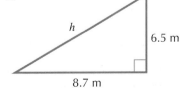

2 Calculate the length of the diagonal AC in the rectangle ABCD.

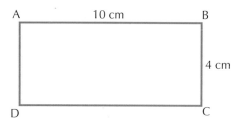

Give your answer correct to one decimal place.

3 Calculate the length of the diagonal of a square with side length 5 cm. Give your answer correct to one decimal place.

4 In triangle XYZ, XY = 7 cm and YZ = 10 cm. There is a right angle at Y.

Calculate the length of XZ. Give your answer correct to one decimal place.

5 The triangle PQR is drawn on a coordinate grid, as shown in the diagram.

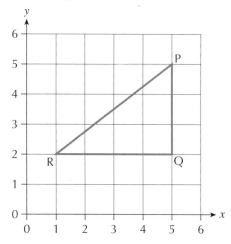

a The length of PQ is 3 units. Write down the length of QR.

b Calculate the length of PR.

6 A ship sails due east from Port A to Port B.

It then sails due north to Port C.

Finally, it sails directly back to Port A.

How far does it sail, from Port C to Port A? Give your answer correct to one decimal place.

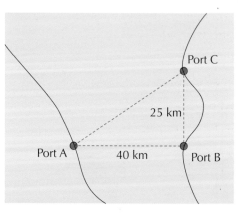

 7 Calculate the length between each pair of coordinate points.

a A(2, 3) and B(4, 7) **b** C(1, 5) and D(4, 3)

c E(−1, 0) and F(2, −3) **d** G(−5, 1) and H(4, −3)

A Calculate the lengths labelled x, y and z in the diagram.

Give your answers correct to one decimal place.

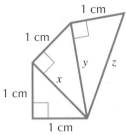

B **a** Now repeat question **A**, but leave your answers as square roots.

Is there a pattern?

b Suppose more right-angled triangles are added to the diagram.

Write down the lengths of the hypotenuse for the next four right-angled triangles in the sequence.

c The diagram is composed of n right-angled triangles. What is the length of the hypotenuse of the nth triangle?

6.3 Calculating the length of a shorter side

Learning objectives

- To calculate the length of a shorter side in a right-angled triangle
- To show that a triangle is right-angled

Key word

shorter side

Any right-angled triangle has a hypotenuse (opposite the right angle) and two **shorter sides**.

You can use Pythagoras' theorem to calculate the length of a shorter side in a right-angled triangle, but will need to use subtraction.

Pythagoras' theorem is:

$$c^2 = a^2 + b^2$$

where c is the hypotenuse and a and b are the shorter sides.

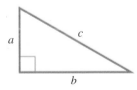

You can rearrange the formula to obtain the form you need.

$$a^2 = c^2 - b^2 \text{ or } b^2 = c^2 - a^2$$

Use this version of Pythagoras' theorem to calculate the length of a shorter side, when you know the hypotenuse and another side.

Example 3

Calculate the length labelled x in this triangle.

The side labelled x is a shorter side.

Using Pythagoras' theorem:

$$x^2 = 9^2 - 7^2$$
$$= 81 - 49$$
$$= 32$$

So $x = \sqrt{32} = 5.7$ cm (1 dp).

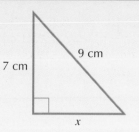

Try this sequence of keystrokes on your calculator.

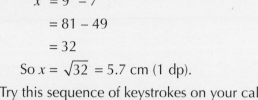

Example 4

Show that triangle ABC is a right-angled triangle.

$$AC^2 = 4^2 = 16$$
$$2.4^2 + 3.2^2 = 5.76 + 10.24 = 16$$

So the triangle has a right angle at B.

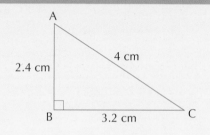

Exercise 6C

1 Calculate the length of the unknown shorter side in each right-angled triangle. Give your answers correct to one decimal place.

a

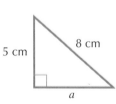

b

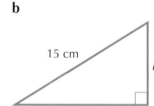

c

d

e

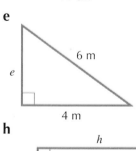

f

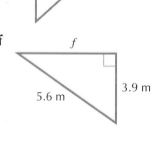

g

h

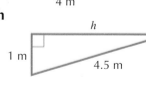

2 In triangle ABC, AC = 15 cm and AB = 10 cm. The angle at B is a right angle.
Calculate the length of BC.
Give your answer correct to one decimal place.

3 Calculate the length of CD on this diagram.
Give your answer correct to one decimal place.

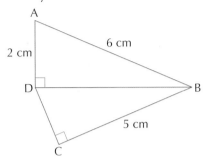

4 An equilateral triangle has sides of length 10 cm.
   a Calculate the perpendicular height, h, of the triangle.
   b Calculate the area of the triangle.
   Give your answers correct to one decimal place.

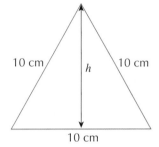

5 Show that these are all right-angled triangles.

   a

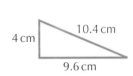

   b

   c

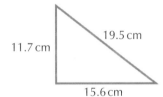

MR **6** A radio mast 5.25 m tall is anchored to the ground by a cable that is 8.75 m long.
The cable is anchored to a point 7 m from the base of the mast.
Is the mast vertical? Explain your answer.

PS **7** Calculate the area of triangle ABD.
Give your answer correct to one decimal place.

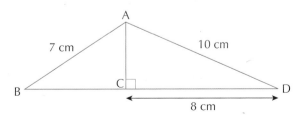

Activity: Pythagoras and angles

The table shows the lengths, in centimetres, of the three sides of six triangles.

A Construct each triangle accurately, using a ruler and a pair of compasses.

Label each one as in this triangle.

B Copy and complete the table.

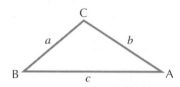

a	b	c	a^2	b^2	c^2	$a^2 + b^2$	Is $a^2 + b^2 = c^2$? Is $a^2 + b^2 > c^2$? Is $a^2 + b^2 < c^2$? Write =, > or <	Is the angle at C a right angle, acute or obtuse?
3	4	5						
4	5	7						
5	6	7						
5	12	13						
4	8	10						
7	8	9						

C Use the results in the last two columns of your table to write down a rule.

D Test your rule by drawing some more triangles with different lengths for a, b and c.

6.4 Using Pythagoras' theorem to solve problems

Learning objective

- To use Pythagoras' theorem to solve problems

You can use Pythagoras' theorem to solve various practical problems in 2D.

Follow these steps.

- Draw a diagram for the problem, clearly showing the right angle.
- Decide whether you need to find the hypotenuse or one of the shorter sides.
- Label the unknown side x.
- Use Pythagoras' theorem to calculate the value of x.
- Round your answer to a suitable degree of accuracy.

Example 5

A ship sails 4 km due east. It then sails for a further 5 km due south. Calculate the distance the ship would have travelled, if it had sailed directly from its starting point to its finishing point.

First, draw a diagram to show the distances sailed by the ship.

Then label the direct distance, x.

Now use Pythagoras' theorem to calculate the length of the hypotenuse.

$$x^2 = 4^2 + 5^2$$
$$= 16 + 25$$
$$= 41$$

So $x = \sqrt{41} = 6.4$ km (1 dp).

Exercise 6D ▦

In this exercise, give your answers to a suitable degree of accuracy.

1 An aircraft flies 80 km due north.

Then it flies 72 km due west. Calculate how far the aircraft would have travelled if it had travelled directly from its starting point to its destination.

2 A flagpole is 10 m high. It is held in position by four ropes that are each fixed to the ground, 4 m away from the foot of the flagpole, to make a square. The diagram shows the flagpole and two ropes that are opposite each other.

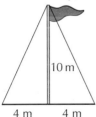

Calculate the length of each rope.

3 A hockey pitch measures 90 m by 55 m.

Calculate the length of a diagonal of the pitch.

4 An 8 m ladder is placed against a wall so that the foot of the ladder is 2 m away from the bottom of the wall.

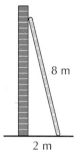

8 m

2 m

Calculate how far the ladder reaches up the wall.

5 The lengths of the sides of an equilateral triangle are 10 cm.
Calculate the perpendicular height of the triangle.

6 The diagram shows the side wall of a shed.
Calculate the length of the sloping roof.

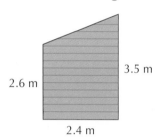

3.5 m

2.6 m

2.4 m

7 The diagram shows a walkway leading from the dock on to a ferry.
Calculate the vertical height of the top of the walkway above the dock.

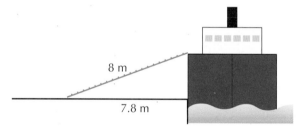

8 m

7.8 m

MR **8** The lengths of two sides of a right-angled triangle are 20 cm and 30 cm. Calculate the length of the third side, if it is:

a the hypotenuse **b** not the hypotenuse.

9 ABC is an isosceles triangle.

a Calculate the perpendicular height, h.

b Hence calculate the area of the triangle.

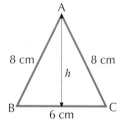

A

8 cm 8 cm

h

B
6 cm
C

10 The diagram shows three towns X, Y and Z connected by straight roads.

Calculate the shortest distance from X to the road connecting Y and Z.

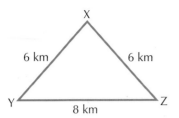

11 In rectangle ABCD, AB = 20 cm and CD = 16 cm.

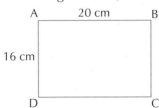

E is a point on AB and AE : EB = 2 : 3.

F is a point on BC and BF : FC = 3 : 1.

Calculate the length of EF.

Problem solving: Pythagorean triples

Any set of numbers that obey the rule:

$$c^2 = a^2 + b^2$$

where a, b and c are whole numbers is called a **Pythagorean triple**.

In this table, the numbers in each row form a Pythagorean triple, but with the extra condition that a is an odd number.

a	b	c
3	4	5
5	12	13
7	24	25

A Continue the table to find other Pythagorean triples, ensuring that a is an odd number in each case.

You could use a spreadsheet to help you.

B Work out the formula that Pythagoras discovered, giving b and c when the value of a is known.

C Try to find out whether multiples of any Pythagorean triple still give another Pythagorean triple.

Ready to progress?

Review questions

1 a Calculate the length of the hypotenuse of this right-angled triangle.

Give your answer correct to one decimal place.

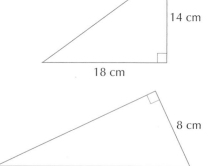

14 cm

18 cm

b Calculate the length of the shorter side of this right-angled triangle.

Give your answer correct to one decimal place.

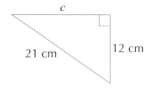

8 cm

15 cm

2 Calculate the length of the unknown side in each of these right-angled triangles.

Give your answers correct to one decimal place.

a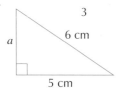

3

6 cm

a

5 cm

b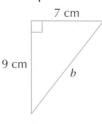

7 cm

9 cm

b

c

c

21 cm

12 cm

d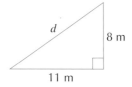

d

8 m

11 m

e

e

3.2 m

1 m

f

8.5 m

10.8 m

f

3 Four right-angled triangles are cut from a rectangular piece of card to make a pentagon.

a Write down the perimeter of the original card.

b Work out the perimeter of the pentagon.

c Work out the perimeter of the pentagon as a percentage of the perimeter of the card.

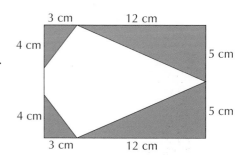

3 cm   12 cm

4 cm

5 cm

4 cm

5 cm

3 cm   12 cm

PS 4 Two right-angled triangles QRS and PQR are joined together to make a larger triangle, SRP.

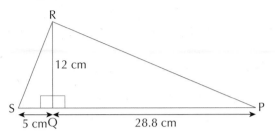

a Show that the perimeter of triangle SRP is 78 cm.

b Show that triangle SRP is also a right-angled triangle.

MR 5 Simon draws a quadrilateral on a sheet of paper. Its side lengths are all 8.2 cm and the diagonal is 12 cm.

Simon claims that it is a square. Is he correct?

Explain your answer.

MR 6 A garden shed has these measurements.

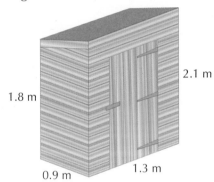

Calculate the area of the sloping roof.

Give your answer correct to two decimal places.

PS 7 The sides of a regular hexagon are all 6 cm long.

Calculate the length of the diagonal, marked x on the diagram.

Give your answer correct to one decimal place.

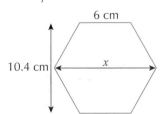

Activity
Practical Pythagoras

This is a practical demonstration to show Pythagoras' theorem. You will need a pencil, a ruler, a sheet of A4 paper, a sheet of thin card and a pair of scissors.

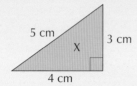

- In the centre of your paper, draw the right-angled triangle X, as shown.

- On the card, draw eight more triangles identical to X. Cut them out and place them to one side.

- On your original triangle, X, draw squares on each of the three sides. Label them A, B and C, as shown.

ΠΥΘΑΓΟΡΑΣ Ο ΣΑΜΙΟΣ
580 - 496 π.χ

- On the card, draw another diagram identical to this.
  Cut out the squares A, B and C.

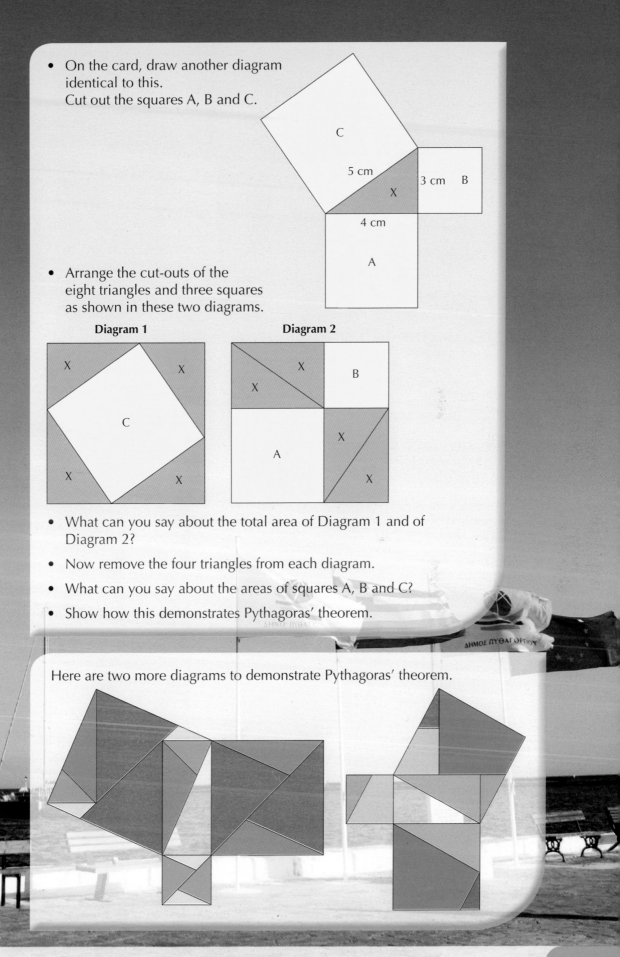

C

5 cm

3 cm   B

X

4 cm

A

- Arrange the cut-outs of the eight triangles and three squares as shown in these two diagrams.

Diagram 1

X   X

C

X   X

Diagram 2

X

X

B

X

A

X

- What can you say about the total area of Diagram 1 and of Diagram 2?

- Now remove the four triangles from each diagram.

- What can you say about the areas of squares A, B and C?

- Show how this demonstrates Pythagoras' theorem.

Here are two more diagrams to demonstrate Pythagoras' theorem.

7

Fractions

This chapter is going to show you:

- how to multiply any two fractions or mixed numbers
- how to divide any two fractions or mixed numbers.

You should already know:

- how to add or subtract two fractions.

About this chapter

If you could count all the pips on this board, you would see that the fraction of the square that is green is $\frac{36}{121}$ or about 0.30.

The first value is exact. The second is a decimal approximation. A more accurate approximation is 0.298.

You can write numbers that are not integers in two different ways: as decimals or as fractions. But why do you need two different ways to write numbers?

In many cases, such as money or measuring, decimals are the better choice. However, some simple fractions, such as one third, cannot be written down exactly as decimals. In those cases it may be better to use fractions.

In other cases, such as common formulae involving simple fractions, writing them with decimals would make them more difficult to remember. It would not be easy to calculate accurately. An example is the volume of a sphere, which you will meet later in your mathematics course. It involves the fraction $\frac{4}{3}$ which, as a decimal, is 1.333 333 .... In this case, it is easier to write the number as a fraction.

7.1 Adding and subtracting fractions

Learning objective

- To add or subtract any two mixed numbers confidently

In this section, you will review the addition and subtraction of fractions, including mixed numbers. You need to be able to do this without using a calculator.

Before you can add or subtract two fractions you need to make sure they are written with the same number in the denominator. This example will illustrate the method.

Example 1

Work these out.   **a** $4\frac{1}{2} + 2\frac{2}{3}$ **b** $3\frac{1}{4} - \frac{7}{12}$

a $4\frac{1}{2} + 2\frac{2}{3} = \frac{9}{2} + \frac{8}{3}$ Write the mixed numbers as improper fractions.

$\quad \frac{9}{2} + \frac{8}{3} = \frac{27}{6} + \frac{16}{6}$ 6 is a multiple of 2 and 3 so change both fractions to sixths.

$\quad\quad\quad\quad = \frac{43}{6}$ Add the numerators, keep the same denominator.

$\quad\quad\quad\quad = 7\frac{1}{6}$ $43 \div 6 = 7$ remainder 1

b $3\frac{1}{4} - \frac{7}{12} = \frac{13}{4} - \frac{7}{12}$

$\quad \frac{39}{12} - \frac{7}{12} = \frac{32}{12}$ Change $\frac{13}{4}$ to twelfths and then subtract.

$\quad\quad\quad\quad = \frac{8}{3}$ Divide both 32 and 12 by 4 to find an equivalent fraction.

$\quad\quad\quad\quad = 2\frac{2}{3}$

Exercise 7A

1 Work these out.

   a $\frac{3}{4} + \frac{1}{8}$ **b** $\frac{3}{4} + \frac{1}{6}$ **c** $\frac{3}{4} + \frac{2}{3}$ **d** $\frac{3}{4} + \frac{7}{10}$

   e $\frac{3}{4} - \frac{1}{8}$ **f** $\frac{3}{4} - \frac{1}{6}$ **g** $\frac{3}{4} - \frac{2}{3}$ **h** $\frac{3}{4} - \frac{7}{10}$

2 Work these out.

   a $1\frac{1}{2} + 2\frac{3}{4}$ **b** $\frac{7}{8} + 2\frac{1}{4}$ **c** $3\frac{2}{3} + 1\frac{1}{6}$ **d** $3\frac{3}{4} + \frac{5}{6}$

   e $1\frac{1}{4} + \frac{5}{8}$ **f** $4\frac{1}{4} + 1\frac{1}{3}$ **g** $3\frac{3}{4} + 1\frac{5}{6}$ **h** $6\frac{1}{4} + 2\frac{2}{3}$

3 Work these out.

   a $4\frac{1}{2} - \frac{3}{4}$ **b** $2\frac{7}{8} - 2\frac{1}{4}$ **c** $3 - 1\frac{5}{6}$ **d** $3\frac{3}{4} - \frac{2}{3}$

   e $3\frac{1}{4} - 1\frac{5}{8}$ **f** $6\frac{1}{4} - 1\frac{1}{2}$ **g** $4\frac{1}{4} - 1\frac{5}{8}$ **h** $6\frac{3}{4} - 6\frac{2}{3}$

4 Here are two fractions.

$$\frac{7}{12} \qquad \frac{3}{8}$$

Work out:

a the sum of the two fractions

b the difference between the two fractions.

5 Work out the perimeter of this triangle.

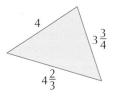

6 **a** Work out the difference between the length and the width of this rectangle.

b Work out the perimeter of the rectangle.

$8\frac{1}{3}$

$4\frac{1}{2}$

PS **7** Here are four numbers.

$$2\frac{1}{3} \qquad 2\frac{1}{6} \qquad 2\frac{1}{4} \qquad 2\frac{5}{12}$$

a Which two fractions have a sum of $4\frac{3}{4}$?

b Which pairs of fractions have a difference of $\frac{1}{12}$?

8 Work these out.

a $1\frac{1}{2} + 2\frac{3}{4} + 3\frac{5}{8}$ **b** $2\frac{3}{4} + 1\frac{1}{2} + 3\frac{1}{3}$ **c** $2\frac{2}{3} + 1\frac{1}{6} + 3\frac{5}{9}$

PS **9** Look at this sum.

$$3\frac{5}{6} + 1\frac{2}{9} = 5\frac{1}{18}$$

Use the sum to write down the answer to each of these calculations.

a $5\frac{1}{18} - 1\frac{2}{9}$ **b** $4\frac{5}{6} + 2\frac{2}{9}$ **c** $1\frac{5}{6} + 6\frac{2}{9}$ **d** $5\frac{1}{18} - 3\frac{2}{9}$

PS **10** Work out the missing numbers in these calculations.

a $4\frac{1}{3} + \ldots = 8$ **b** $2\frac{3}{10} + \ldots = 5\frac{9}{10}$ **c** $1\frac{2}{3} + \ldots = 4\frac{1}{2}$

11 $x = 3\frac{3}{4}$ and $y = 2\frac{2}{5}$

Work these out.

a $x + y$ **b** $x - y$ **c** $2x + y$ **d** $2x - y$

Challenge: Fraction sequence

Here are three numbers that form a linear sequence.

$$1\frac{1}{3} \qquad 3\frac{7}{12} \qquad 5\frac{5}{6}$$

A Work out the difference between the first and second numbers.

B Work out the difference between the second and third numbers.

C Work out the next three numbers in the sequence.

7.2 Multiplying fractions

Learning objective

• To multiply two fractions

To multiply a fraction by an integer, you just multiply the numerator by the integer.

For example:

• $5 \times \frac{2}{3} = \frac{10}{3} = 3\frac{1}{3}$

• $\frac{3}{4}$ of $7 = \frac{3}{4} \times 7 = \frac{21}{4} = 5\frac{1}{4}$

To multiply two fractions, you multiply the numerators and multiply the denominators.

Example 2

Work these out.   **a** $\frac{3}{4}$ of $\frac{1}{2}$ **b** $\frac{2}{3} \times \frac{3}{5}$

a $\frac{3}{4}$ of $\frac{1}{2} = \frac{3}{4} \times \frac{1}{2}$

$= \frac{3 \times 1}{4 \times 2}$ The numerator is 3×1 and the denominator is 4×2.

$= \frac{3}{8}$

b $\frac{2}{3} \times \frac{3}{5} = \frac{6}{15}$ $2 \times 3 = 6$ and $3 \times 5 = 15$

$= \frac{2}{5}$ Simplify the fraction as much as possible.

These rules are just an extension of the method for integers.

5 can be written as $\frac{5}{1}$ so:

$5 \times \frac{2}{3} = \frac{5}{1} \times \frac{2}{3} = \frac{10}{3} = 3\frac{1}{3}$

as before.

Exercise 7B

1 Work these out.

a $2 \times \frac{3}{8}$ **b** $3 \times \frac{3}{4}$ **c** $3 \times \frac{4}{5}$ **d** $4 \times \frac{3}{8}$

2 Work these out.

a $\frac{1}{5}$ of 3   **b** $\frac{4}{5}$ of 3   **c** $\frac{2}{3}$ of 4   **d** $\frac{5}{6}$ of 2

3 Work these out.

a $\frac{1}{2} \times \frac{1}{3}$ b $\frac{1}{2} \times \frac{3}{4}$ c $\frac{1}{2} \times \frac{3}{5}$ d $\frac{1}{2} \times \frac{4}{5}$

e $\frac{1}{3} \times \frac{1}{4}$ f $\frac{1}{3} \times \frac{3}{5}$ g $\frac{1}{3} \times \frac{5}{8}$ h $\frac{1}{3} \times \frac{2}{3}$

4 Work out $\frac{2}{3}$ of:

a $\frac{1}{2}$ b $\frac{3}{4}$ c $\frac{2}{3}$

d $\frac{4}{5}$ e $\frac{1}{8}$ f $\frac{5}{6}$.

5 Work out these multiplications.

a $\frac{2}{3} \times \frac{1}{4}$ b $\frac{3}{5} \times \frac{3}{4}$ c $\frac{5}{8} \times \frac{2}{3}$ d $\frac{3}{8} \times \frac{2}{5}$

e $\frac{2}{5} \times \frac{3}{4}$ f $\frac{3}{8} \times \frac{2}{3}$ g $\frac{4}{9} \times \frac{3}{8}$ h $\frac{4}{5} \times \frac{5}{12}$

6 Work these out.

a $\frac{1}{2} \times \frac{1}{2}$ b $\frac{2}{3} \times \frac{2}{3}$ c $\frac{3}{5} \times \frac{3}{5}$ d $\frac{3}{4} \times \frac{3}{4}$

7 Work these out.

a $\frac{5}{8} \times \frac{4}{5}$ b $\frac{1}{10} \times \frac{5}{6}$ c $\frac{7}{8} \times \frac{2}{3}$ d $\frac{3}{8} \times \frac{2}{5}$

(PS) **8** $\frac{1}{2} \times \frac{a}{b} = \frac{3}{8}$

Work out the values of a and b.

9 Work these out.

a $\left(\frac{1}{4}\right)^2$ b $\left(\frac{3}{4}\right)^2$ c $\left(\frac{2}{3}\right)^2$ d $\left(\frac{4}{5}\right)^2$

10 Work these out.

a $\left(\frac{1}{2} + \frac{1}{4}\right) \times \frac{3}{5}$ b $\left(\frac{2}{3} + \frac{1}{6}\right) \times \frac{3}{10}$ c $\left(\frac{1}{2} - \frac{1}{6}\right) \times \frac{3}{4}$

(MR) **11** Here is a multiplication.

$$\frac{5}{12} \times \frac{4}{15} = \frac{1}{9}$$

Use this result to work these out.

a $\frac{5}{12} \times \frac{8}{15}$ b $\frac{5}{12} \times \frac{2}{15}$ c $\frac{5}{6} \times \frac{4}{15}$

12 $s = \frac{5}{8}$ and $t = \frac{2}{3}$

Work these out.

a st b ts c s^2 d t^2

Investigation: Multiplication diagrams

This diagram represents $3\frac{1}{2}$.

Each rectangle is made up of eight squares and represents one whole.

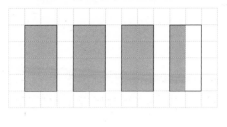

This diagram shows $3\frac{1}{2}$ divided into four equal parts by horizontal lines.

A Explain how the part coloured blue shows that $\frac{1}{4}$ of $3\frac{1}{2} = \frac{7}{8}$.

B Explain what the part coloured green shows.

C Draw a diagram to show that $\frac{1}{5}$ of $3\frac{1}{2} = \frac{7}{10}$.

D Use your diagram to show the value of $\frac{4}{5}$ of $3\frac{1}{2}$.

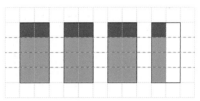

 Hint This time, draw rectangles with ten squares.

7.3 Multiplying mixed numbers

Learning objective

• To multiply one mixed number by another

To multiply two mixed numbers, you need first to change them to improper fractions. Then you can use the method described in the previous section.

Example 3

Work these out.   **a** $\frac{2}{3}$ of $3\frac{1}{2}$ **b** $1\frac{3}{4} \times 2\frac{1}{2}$

a $\frac{2}{3}$ of $3\frac{1}{2} = \frac{2}{3} \times \frac{7}{2}$ Change $3\frac{1}{2}$ to an improper fraction.

$\quad = \frac{14}{6}$ $2 \times 7 = 14$ and $3 \times 2 = 6$

$\quad = \frac{7}{3}$ Simplify the fraction by dividing top and bottom by 2.

$\quad = 2\frac{1}{3}$ Convert the answer to a mixed number.

b $1\frac{3}{4} \times 2\frac{1}{2} = \frac{7}{4} \times \frac{5}{2}$ Change both mixed numbers to improper fractions.

$\quad = \frac{35}{8}$

$\quad = 4\frac{3}{8}$ Convert the answer to a mixed number.

Exercise 7C

1 Work these out.

   a $\frac{1}{2} \times 1\frac{1}{2}$ **b** $\frac{1}{2} \times 1\frac{1}{4}$ **c** $\frac{1}{4} \times 2\frac{1}{2}$ **d** $\frac{1}{4} \times 3\frac{1}{2}$

2 Work these out.

   a $4\frac{1}{2} \times \frac{1}{3}$ **b** $3\frac{1}{2} \times \frac{2}{3}$ **c** $1\frac{1}{4} \times \frac{1}{4}$ **d** $5\frac{1}{2} \times \frac{2}{3}$

(PS) **3**   **a**   Work these out.

      i $\frac{1}{5} \times 1\frac{1}{2}$ **ii** $\frac{1}{5} \times 2\frac{1}{2}$ **iii** $\frac{1}{5} \times 3\frac{1}{2}$ **iv** $\frac{1}{5} \times 4\frac{1}{2}$

     b   The multiplications in part **a** follow a pattern.

        Write down and calculate the next two terms in the pattern.

4 Work these out.

   a $1\frac{1}{3} \times 2\frac{1}{2}$ **b** $1\frac{1}{4} \times 3\frac{1}{2}$ **c** $2\frac{1}{3} \times 2\frac{1}{2}$ **d** $4\frac{1}{2} \times 2\frac{2}{3}$

5 Multiply $2\frac{2}{3}$ by:

   a $1\frac{2}{3}$ **b** $2\frac{1}{3}$ **c** $1\frac{3}{4}$ **d** $5\frac{1}{2}$.

6 Multiply $3\frac{3}{4}$ by:

   a $\frac{2}{5}$ **b** $2\frac{2}{3}$ **c** $1\frac{3}{5}$ **d** $5\frac{1}{3}$.

7 Work these out.

   a $\left(\frac{3}{4}\right)^2$ **b** $\left(2\frac{1}{4}\right)^2$ **c** $\left(4\frac{1}{2}\right)^2$ **d** $\left(1\frac{3}{4}\right)^2$

8   **a**   Work these out.

      i $1\frac{1}{2} \times \frac{2}{3}$ **ii** $1\frac{1}{3} \times \frac{3}{4}$ **iii** $2\frac{1}{2} \times \frac{2}{5}$ **iv** $1\frac{3}{4} \times \frac{4}{7}$

(PS)   **b**   Write down two more multiplications like those in part **a**.

9 Work these out.

   a $1\frac{1}{2} \times 2\frac{1}{2}$ **b** $1\frac{2}{3} \times 2\frac{2}{3}$ **c** $1\frac{3}{4} \times 2\frac{3}{4}$ **d** $1\frac{4}{5} \times 2\frac{4}{5}$

10   **a**   Work out the value of x^2 when x is equal to:

      i $1\frac{1}{2}$ **ii** $1\frac{1}{3}$ **iii** $1\frac{1}{4}$ **iv** $1\frac{2}{5}$.

     b   Which answer in part **a** is closest to 2?

Investigation: Constant perimeter

Here are three rectangles. All lengths are in centimetres.

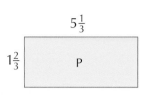

$5\frac{1}{3}$

$1\frac{2}{3}$ P

$4\frac{1}{2}$

$2\frac{1}{2}$ Q

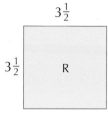

$3\frac{1}{2}$

$3\frac{1}{2}$ R

A Show that each rectangle has the same perimeter.

B Calculate the area of each rectangle.

C Another rectangle has the same perimeter. It is $5\frac{1}{2}$ cm long.
Work out:
   a the width of the rectangle   **b** the area of the rectangle.

D **a** Work out the length and width of another rectangle with the same perimeter as in question C.
   b Work out its area.

E What is special about the shape of the largest rectangle?

7.4 Dividing fractions and mixed numbers

Learning objective

- To divide one fraction or mixed number by another

You already know how to divide a fraction by an integer.
For example:

$$5\frac{1}{2} \div 4 = \frac{11}{2} \div 4 = \frac{11}{8} = 1\frac{3}{8}$$

In this case you multiply the denominator by 4.

To divide by a fraction, you **invert** it or 'turn it upside down'. This means that you swap the numerator and the denominator. You then multiply by the new fraction. Some examples should make the method clear.

Example 4

Work out   **a** $3 \div \frac{2}{3}$ **b** $\frac{2}{3} \div \frac{3}{4}$

a $3 \div \frac{2}{3} = \frac{3}{1} \times \frac{3}{2}$ Write 3 as $\frac{3}{1}$ and invert $\frac{2}{3}$ to get $\frac{3}{2}$.

$= \frac{3 \times 3}{1 \times 2} = \frac{9}{2} = 4\frac{1}{2}$

b $\frac{2}{3} \div \frac{3}{4} = \frac{2}{3} \times \frac{4}{3}$ Invert $\frac{3}{4}$ but do not change $\frac{2}{3}$.

$= \frac{8}{9}$

If the calculation includes mixed numbers, change them to improper fractions first.

Example 5

Work these out.   **a** $2\frac{1}{4} \div 1\frac{1}{3}$ **b** $1\frac{1}{2} \div 3\frac{1}{2}$

$\quad$ **a** $\quad 2\frac{1}{4} \div 1\frac{1}{3} = \frac{9}{4} \div \frac{4}{3}$ $\qquad$ Write the mixed numbers as improper fractions.

$\qquad\qquad = \frac{9}{4} \times \frac{3}{4}$ $\qquad\qquad$ Invert $\frac{4}{3}$ and multiply.

$\qquad\qquad = \frac{27}{16}$

$\qquad\qquad = 1\frac{11}{16}$ $\qquad\qquad$ Write the improper fraction as a mixed number.

$\quad$ **b** $\quad 1\frac{1}{2} \div 3\frac{1}{2} = \frac{3}{2} \div \frac{7}{2}$ $\qquad$ Write the mixed number as improper fractions.

$\qquad\qquad = \frac{3}{2} \times \frac{2}{7}$ $\qquad\qquad$ Invert $\frac{7}{2}$ and multiply.

$\qquad\qquad = \frac{6}{14}$

$\qquad\qquad = \frac{3}{7}$ $\qquad\qquad$ Simplify the fraction as much as possible.

Exercise 7D

1 Work these out.

$\quad$ **a** $\quad 3 \div \frac{1}{2}$ $\qquad\qquad$ **b** $\quad 3 \div \frac{1}{3}$ $\qquad\qquad$ **c** $\quad 3 \div \frac{2}{3}$ $\qquad\qquad$ **d** $\quad 3 \div \frac{1}{4}$

$\quad$ **e** $\quad 3 \div \frac{3}{4}$ $\qquad\qquad$ **f** $\quad 3 \div \frac{1}{5}$ $\qquad\qquad$ **g** $\quad 3 \div \frac{2}{5}$ $\qquad\qquad$ **h** $\quad 3 \div \frac{3}{5}$

2 Work these out.

$\quad$ **a** $\quad 2 \div \frac{1}{4}$ $\qquad\qquad$ **b** $\quad 5 \div \frac{2}{3}$ $\qquad\qquad$ **c** $\quad 8 \div \frac{3}{4}$ $\qquad\qquad$ **d** $\quad 6 \div \frac{4}{5}$

$\quad$ **e** $\quad 10 \div \frac{4}{5}$ $\qquad\qquad$ **f** $\quad 3 \div \frac{9}{10}$ $\qquad\qquad$ **g** $\quad 12 \div \frac{3}{4}$ $\qquad\qquad$ **h** $\quad 2 \div \frac{5}{12}$

3 Work these out.

$\quad$ **a** $\quad \frac{1}{2} \div \frac{1}{4}$ $\qquad\qquad$ **b** $\quad \frac{1}{2} \div \frac{1}{3}$ $\qquad\qquad$ **c** $\quad \frac{1}{2} \div \frac{1}{5}$ $\qquad\qquad$ **d** $\quad \frac{1}{2} \div \frac{1}{8}$

4 Work out the following divisions.

$\quad$ **a** $\quad \frac{1}{2} \div \frac{2}{5}$ $\qquad\qquad$ **b** $\quad \frac{1}{4} \div \frac{3}{4}$ $\qquad\qquad$ **c** $\quad \frac{2}{5} \div \frac{3}{10}$ $\qquad\qquad$ **d** $\quad \frac{1}{6} \div \frac{2}{3}$

5 **a** Work these out.

$\quad\quad$ **i** $\quad \frac{1}{4} \div \frac{1}{3}$ $\qquad$ **ii** $\quad \frac{1}{3} \div \frac{1}{4}$ $\qquad$ **iii** $\quad \frac{1}{3} \div \frac{1}{6}$ $\qquad$ **iv** $\quad \frac{1}{6} \div \frac{1}{3}$

$\quad\quad$ **v** $\quad \frac{1}{2} \div \frac{2}{3}$ $\qquad$ **vi** $\quad \frac{2}{3} \div \frac{1}{2}$ $\qquad$ **vii** $\quad \frac{3}{4} \div \frac{2}{3}$ $\qquad$ **viii** $\quad \frac{2}{3} \div \frac{3}{4}$

$\quad$ **b** What do you notice about the pairs of answers in part **a**?

6 Work these out.

  a $2\frac{1}{2} \div \frac{1}{2}$ **b** $2\frac{1}{2} \div \frac{1}{4}$ **c** $2\frac{1}{2} \div \frac{3}{4}$ **d** $2\frac{1}{2} \div \frac{5}{8}$

7 Work these out.

  a $1\frac{1}{2} \div \frac{2}{3}$ **b** $3\frac{1}{4} \div \frac{3}{4}$ **c** $4\frac{1}{2} \div \frac{2}{3}$ **d** $7\frac{1}{2} \div \frac{3}{5}$

 8 **a** Work out $2\frac{1}{4} \div \frac{3}{8}$.

  b Check your calculation in part **a** by multiplying the answer by $\frac{3}{8}$.

 9 **a** Calculate $2\frac{1}{2} \div \frac{4}{5}$.

  b Check your calculation in part **a** by multiplying the answer by $\frac{4}{5}$.

10 Calculate:

  a $4\frac{1}{2} \div 1\frac{1}{2}$ **b** $3\frac{3}{4} \div 2\frac{1}{2}$ **c** $6\frac{1}{2} \div 1\frac{1}{2}$ **d** $8\frac{1}{2} \div 3\frac{1}{2}$

  e $1\frac{1}{2} \div 4\frac{1}{2}$ **f** $2\frac{1}{2} \div 5\frac{1}{2}$ **g** $\frac{1}{2} \div 6\frac{1}{2}$ **h** $3\frac{1}{2} \div 4\frac{1}{2}$.

11 Work these out.

  a $2\frac{1}{2} \div 1\frac{1}{4}$ **b** $3\frac{1}{3} \div 1\frac{1}{2}$ **c** $1\frac{2}{3} \div 1\frac{1}{4}$ **d** $1\frac{1}{2} \div 7\frac{1}{2}$

  e $4\frac{1}{2} \div 6\frac{3}{4}$ **f** $12\frac{1}{2} \div 7\frac{1}{2}$ **g** $3\frac{1}{2} \div 4\frac{2}{3}$ **h** $10\frac{1}{2} \div 4\frac{2}{3}$

 12 **a** Calculate $5\frac{1}{4} \div 2\frac{1}{3}$.

  b Check your calculation in part **a** by multiplying the answer by $2\frac{1}{3}$.

Challenge: Algebra with fractions

$x = 3\frac{1}{2}$ $y = 2\frac{1}{3}$ $z = 1\frac{3}{4}$

Work out the value of each expression.

  A $x + y$ **B** $y - z$ **C** xy **D** $\dfrac{x}{y}$

  E $2z$ **F** z^2 **G** $\dfrac{z}{y}$ **H** $x + y + z$

Ready to progress?

I can add or subtract two mixed numbers.
I can multiply two simple fractions.

I can multiply two mixed numbers.
I can divide two simple fractions.
I can divide two mixed numbers.

Review questions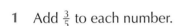

1 Add $\frac{3}{5}$ to each number.

    a $\frac{1}{2}$ b $\frac{3}{10}$ c $\frac{1}{3}$ d $\frac{2}{3}$ e $\frac{3}{4}$

2 Work out the difference between:

    a $\frac{2}{3}$ and $\frac{1}{2}$ b $\frac{1}{4}$ and $\frac{2}{3}$ c $\frac{3}{4}$ and $\frac{2}{3}$ d $\frac{1}{3}$ and $1\frac{1}{2}$.

3 Work these out.

    a $2 \times \frac{5}{8}$ b $3 \times \frac{3}{8}$ c $\frac{2}{3} \times 8$ d $\frac{2}{5} \times 11$

4 Work out $\frac{3}{4}$ of:

    a 2   b $1\frac{1}{2}$ c $1\frac{1}{4}$ d $\frac{3}{4}$.

5 Work these out.

    a $\frac{5}{8} + \frac{1}{2}$ b $\frac{5}{8} - \frac{1}{2}$ c $\frac{5}{8} \times \frac{1}{2}$ d $\frac{5}{8} \div \frac{1}{2}$

6 Work these out.

    a $\frac{2}{3}$ of $\frac{1}{2}$ b $\frac{1}{2}$ of $\frac{2}{3}$ c $\frac{2}{3} \div \frac{1}{2}$ d $\frac{1}{2} \div \frac{2}{3}$

7 Work these out.

    a $2\frac{2}{3} \times \frac{1}{4}$ b $2\frac{1}{4} \times \frac{2}{3}$ c $\frac{3}{5} \times 3\frac{3}{4}$ d $\frac{3}{4} \times 3\frac{3}{5}$

(PS) 8 a Work these out.

      i $4\frac{1}{2} \times \frac{1}{3}$ ii $4\frac{1}{2} \times 1\frac{1}{3}$ iii $4\frac{1}{2} \times 2\frac{1}{3}$ iv $4\frac{1}{2} \times 3\frac{1}{3}$

(MR)   b How will the sequence of questions and answers in part **a** continue?

9 Work out the areas of these rectangles.

a
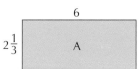
$2\frac{1}{3}$ 6   A

b
$5\frac{1}{2}$ $2\frac{1}{3}$ B

c
$2\frac{2}{5}$ $2\frac{1}{3}$ C

10 Work out $\left(2\frac{1}{2}\right)^2 + \left(3\frac{1}{2}\right)^2$.

11 Work these out.

a $5 \div \frac{1}{4}$ b $\frac{1}{4} \div 5$ c $5 \div \frac{2}{3}$ d $\frac{2}{3} \div 5$

12 Work these out.

a $\frac{1}{3} \div \frac{1}{4}$ b $\frac{1}{5} \div \frac{2}{3}$ c $\frac{5}{8} \div \frac{1}{4}$ d $\frac{7}{8} \div \frac{3}{4}$

13 Calculate:

a $4\frac{1}{6} \div 1\frac{1}{4}$ b $10\frac{1}{2} \div 1\frac{3}{4}$ c $1\frac{1}{8} \div 4\frac{1}{2}$ d $3\frac{2}{3} \div 5\frac{1}{2}$.

14 a   Work out $15\frac{3}{4} \div 2\frac{5}{8}$.

    b   The mass of a suitcase is $15\frac{3}{4}$ kg and the mass of a briefcase is $2\frac{5}{8}$ kg.

       Write the ratio of the mass of the suitcase to the mass of the briefcase as simply as possible.

15 $x = 1\frac{1}{2}$ and $y = 1\frac{2}{3}$.

   Work out the value of each expression.

a $3x$ b $6y$ c $x + y$ d $y - x$

e xy f yx g $x \div y$ h $y \div x$

16 Here is a sequence of numbers.

     4   6   9 $13\frac{1}{2}$

a   Copy and complete this sentence.

     To find the next number you multiply the previous number by …

b   Work out the next number after $13\frac{1}{2}$.

Investigation

Fractions from one to six

1 Here are three fractions.

$$\frac{1}{2} \quad \frac{3}{4} \quad \frac{5}{6}$$

Using two of these fractions and one operation, $+$, $-$, $\times$ or $\div$, you can make other numbers.

For example:

$$\frac{1}{2} + \frac{3}{4} = 1\frac{1}{4}$$

a Show how you could make these numbers.

    i $\frac{5}{12}$ **ii** $\frac{3}{8}$ **iii** $\frac{1}{12}$ **iv** $\frac{3}{5}$ **v** $\frac{1}{3}$ **vi** $1\frac{1}{2}$

b What other positive numbers is it possible to make?

c Explain how you know you have found all the possible answers.

2 Using the digits 1, 2, 3, 4, 5 and 6 you can make two different mixed numbers of the form $a\frac{b}{c}$.

For example, you could make $2\frac{1}{6}$ and $3\frac{4}{5}$.

The fractional part must be less than one and it must be in its lowest terms. You cannot have $3\frac{2}{4}$ or $3\frac{5}{3}$.

  a Write down another possible pair of numbers you could make.

  b The numbers could be multiplied. Show that:
  $$2\frac{1}{6} \times 3\frac{4}{5} = 8\frac{7}{30}$$

  c Work out the product of the two numbers you wrote down in part **a**.

  d Work out the largest possible product you can get by multiplying two numbers in this way.

Algebra

This chapter is going to show you:

- how to expand a bracket when powers are involved
- how to factorise an expression when powers are involved
- how to expand the product of two brackets.

You should already know:

- how to expand and factorise expressions in simple cases.

About this chapter

When the wind catches a sail, or a kite, it takes on a curved shape like those shown on this page. Mathematicians have been interested in the shapes that form naturally, and have used algebra to investigate them. They have used their skills in rearranging expressions and solving equations that involve powers of a variable, such as x^2 or x^3.

These and higher powers occur in many areas of mathematics and science. You have met formulae that involve powers, such as the formulae for the area of a circle and the volume of a cube. The equations of curved lines often involve powers.

In this chapter you will extend your algebraic skills to deal with expressions involving powers.

8.1 More about brackets

Learning objective

* To expand a term with a variable or constant outside brackets

Key word

variable

You already know how to expand (or multiply out) an expression with brackets, when there is a number or a variable on the outside. Here are some examples to remind you.

* $2(x - 5) = 2x - 10$

* $3(2a + 4) = 6a + 12$

Remember that x and a are the **variables** in these examples.

Now look at this example.

* $2x(3x - 5)$

Expanding gives two terms. The first is $2x \times 3x$ and the second is $2x \times (-5)$.

You write $2x \times 3x$ as $6x^2$ because $2 \times 3 = 6$ and $x \times x = x^2$.

You write $2x \times (-5)$ as $-10x$.

So $2x(3x - 5) = 6x^2 - 10x$.

It can be helpful to draw lines to remind you what to multiply by what.

$2x(3x - 5)$

Example 1

Expand these brackets.

a $2x(x + 8)$ **b** $3a(a + 4b)$ **c** $3t(5 - 2t)$ **d** $2x(5x - 3)$

a $2x(x + 8) = 2x^2 + 16x$ $2x \times x = 2x^2$ and $2x \times 8 = 16x$

b $3a(a + 4b) = 3a^2 + 12ab$ $3a \times a = 3a^2$ and $3a \times 4b = 12ab$

c $3t(5 - 2t) = 15t - 6t^2$ $3t \times 5 = 15t$ and $3t \times (-2t) = -6t^2$

d $2x(5x - 3) = 10x^2 - 6x$ $2x \times 5x = 10x^2$ and $2x \times (-3) = -6x$

Exercise 8A

1 Expand the brackets.

    a $4(x + 1)$ **b** $3(y - 7)$ **c** $8(2 - x)$ **d** $4(5 - t)$

2 Expand the brackets.

    a $4(x + 5)$ **b** $x(x + 5)$ **c** $7(y + 10)$ **d** $t(t - 6)$

    e $7(n - 9)$ **f** $k(k - 3)$ **g** $x(5 + x)$ **h** $y(10 - y)$

3 Expand the brackets.

   a $2(x + y)$ **b** $x(x + 2)$ **c** $x(x + y)$ **d** $t(t - u)$

   e $n(4 - n)$ **f** $t(r - t)$ **g** $x(y - x)$ **h** $m(m - 7)$

4 Expand the brackets.

   a $2(2x + 1)$ **b** $4(3a - b)$ **c** $x(2x + 7y)$ **d** $x(3 - 5y)$

   e $2(2a - 3b)$ **f** $2x(x - 8)$ **g** $3x(x - z)$ **h** $x(3x + 2y)$

 5 Write down an expression for the area of:

   a square A

   b rectangle B

   c the whole shape.

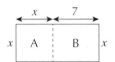

 6 Write an expression for the area of each rectangle. Remember to use brackets.
Then expand the brackets.

   a   **b**   **c**

7 Expand the brackets and simplify as much as possible.

   a $x(x + 4) + 3x$ **b** $x(x - 3) - x$ **c** $x(x + 7) - 5x$ **d** $x(4 + x) - 3x$

8 Simplify each expression.

   a $2x \times 3x$ **b** $2t \times 5t$ **c** $4a \times 6a$ **d** $3x \times x$ **e** $5a \times 2b$

   f $5x \times 3y$ **g** $4n \times 2t$ **h** $6x \times 6x$ **i** $(2x)^2$ **j** $(3x)^2$

9 Multiply out the brackets.

   a $2t(t + 5)$ **b** $2t(2t + 5)$ **c** $2t(3t + 5)$ **d** $2t(5x + 3)$

10 Expand the brackets.

   a $2x(4x - 1)$ **b** $3x(2x - 3)$ **c** $2y(8 - 3y)$ **d** $2n(4 - 7n)$

   e $2x(3x + y)$ **f** $4a(2a + 3b)$ **g** $3t(4s - 3t)$ **h** $6t(6t - 1)$

Challenge: Cubes

$x^2 = x \times x$

So $x \times x^2 = x \times x \times x = x^3$

Use this fact to help you to these expand these brackets.

A $x(x^2 + 2)$ **B** $t(t^2 + t)$ **C** $y(y^2 + 3y)$ **D** $k(k^2 - 6k)$

E $x^2(x + 2)$ **F** $t^2(t - 3)$ **G** $2n(n^2 + 1)$ **H** $2t^2(t - 6)$

8.2 Factorising expressions containing powers

Learning objective

- To take out a variable as a factor

When you factorise an expression you take any common factor and write it outside the brackets. Here is an example to remind you.

$8x + 12 = 4(2x + 3)$ In this case, 4 is the highest common factor of 8 and 12.

You could also write $8x + 12 = 2(4x + 6)$ but this has not been factorised as much as possible. It might happen that a variable is a common factor. If so, it can be put outside the brackets.

Example 2

Factorise each expression as much as possible.

a $6xy - 2x$ **b** $t^2 + 6t$ **c** $4x^2 - 10x$

  a $6xy - 2x = 2(3xy - x)$ 2 is a factor of 6 and 2. However, you can factorise further.

                $= 2x(3y - 1)$ x is also a factor of both terms. Take it outside the brackets.

  b $t^2 + 6t = t(t + 6)$ t is a common factor. Check this by expanding the brackets.

  c $4x^2 - 10x = 2x(2x - 5)$ Both 2 and x are common factors.

Exercise 8B

1   Factorise each expression.

     a $4x + 8$ **b** $12y - 15$ **c** $14 - 7x$ **d** $32y + 40$

2   Complete each factorisation.

     a $x^2 - 3x = x(\ldots)$ **b** $t^2 + 5t = t(\ldots)$ **c** $y^2 - 4y = y(\ldots)$ **d** $6n + n^2 = n(\ldots)$

3   Factorise each expression as much as possible.

     a $x^2 + 6x$ **b** $y^2 - 9y$ **c** $z^2 + z$ **d** $2n - n^2$

     e $12t + t^2$ **f** $x^2 - 7x$ **g** $20n + n^2$ **h** $3x - x^2$

4   Factorise each expression as much as possible.

     a $x^2 + kx$ **b** $x^2 - tx$ **c** $x^2 + xy$ **d** $x^2 - 3xy$

     e $ax + x^2$ **f** $2cx + x^2$ **g** $3x^2 + x$ **h** $4n^2 - n$

5   Some of these expressions can be factorised and some cannot.

     Factorise them if you can. Write 'Not possible' if they can't be factorised.

     a $x^2 + 4$ **b** $x^2 - 6x$ **c** $4x - 10$ **d** $7x - x^2$

     e $12x + x^2$ **f** $20 - 10x$ **g** $13x - x^2$ **h** $4x^2 + 1$

6 Factorise each expression as much as possible.

   a $6x + 12$ **b** $12x - 8y$ **c** $10 - 6t$ **d** $2x^2 + 8$

   e $9t^2 - 6$ **f** $10x^2 + 5y^2$ **g** $8a^2 - 12$ **h** $6a + 9c$

7 Complete each factorisation.

   a $6x^2 + 12x = 6x(\ldots)$ **b** $9y + 6y^2 = 3y(\ldots)$ **c** $10k^2 - 25k = 5k(\ldots)$

   d $10n^2 + 12n = \ldots(5n + 6)$ **e** $32a - 40a^2 = \ldots(4 - 5a)$ **f** $24x^2 + 16x = \ldots(3x + 2)$

8 Factorise each expression as much as possible.

   a $4x^2 + 4x$ **b** $6y^2 - 6y$ **c** $2t^2 + 10t$ **d** $8x^2 - 4x$

   e $20a + 30a^2$ **f** $12t - 16t^2$ **g** $18n^2 + 12n$ **h** $6x^2 - 2$

9 Complete these factorisations.

   a $2x^2 + 4xy = 2x(\ldots + \ldots)$ **b** $6a^2 - 9ab = \ldots(\ldots - \ldots)$

   c $12pq - 16q^2 = 4q(\ldots - \ldots)$

10 Factorise each expression as much as possible.

   a $12x^2 + 12xy$ **b** $4a^2 - 8ab$ **c** $5xy + 5xz$ **d** $8mn - 4mp$

   e $3xy + 6y^2$ **f** $10c + 10cd$ **g** $3ab - 6bc$ **h** $2ab + 12b$

11 Factorise each expression as much as possible.

   a $a^3 + 2a$ **b** $x^2 - 2x^3$ **c** $6n + 3n^3$ **d** $4x^3 - 2x^2$

Challenge: Simplifying expressions

In each expression:

a expand the brackets

b collect like terms

c factorise the resulting expression as much as possible.

   A $2(x + 3) + 2(x - 1)$ **B** $4(x + 3) + 2(x - 3)$ **C** $x(x + 1) + x(x + 5)$

   D $x(x + 4) + x(x - 6)$ **E** $2x(x + 3) + x(x - 3)$ **F** $x(2x + 3) + x(x - 6)$

8.3 Expanding the product of two brackets

Learning objective

• To multiply out two brackets

This expression has two brackets:

   $(x + 3)(x - 6)$

To multiply out the brackets. you must multiply each term in the first set of brackets by each term in the second set of brackets.

• You multiply x by x and by -6.   That gives x^2 and $-6x$.

• You multiply 3 by x and by -6.   That gives $3x$ and -18.

So:

$$(x + 3)(x - 6) = x^2 - 6x + 3x - 18$$

Then you can combine the middle two terms (they are like terms) to give:

$$(x + 3)(x - 6) = x^2 - 6x + 3x - 18 = x^2 - 3x - 18$$

Be careful! Remember to do all four multiplications.

You can draw lines to remind you, like this.

$(x + 3)(x - 6)$

They make a 'smiley face'.

Or you can put the terms in a grid, like this.

	x	$+3$
x	x^2	$+3x$
-6	$-6x$	-18

Always collect like terms if you can.

Example 3

Expand the brackets. Write the result as simply as possible.

a $(x - 3)(x - 5)$ **b** $(a + b)(a - c)$

a Here is a grid for $(x - 3)(x - 5)$.

	x	-3
x	x^2	$-3x$
-5	$-5x$	$+15$

$$(x - 3)(x - 5) = x^2 - 3x - 5x + 15 \qquad \text{Notice that } -3 \times -5 = +15.$$
$$= x^2 - 8x + 15 \qquad \text{Combine like terms: } -3x + -5x = -8x.$$

b Here is a 'smiley face' for $(a + b)(a - c)$.

$(a + b)(a - c)$

$$(a + b)(a - c) = a^2 - ac + ba - bc \qquad \text{This cannot be simplified. There are no like terms.}$$

You could write the answer to part **b** in different ways.

You can:

- change the order of the terms
- change the order of the variables in each term.

For example, you could write the answer as:

$$(a + b)(a - c) = a^2 + ab - cb - ca$$

1 Expand the brackets.

   a $(a + 1)(b + 1)$ **b** $(c + 1)(d + 3)$ **c** $(p + 4)(q + 2)$ **d** $(s + 5)(t + 7)$

2 Expand the brackets.

   a $(x - 1)(y + 2)$ **b** $(a - 4)(b + 5)$ **c** $(n - 1)(m - 3)$ **d** $(p + 5)(q - 6)$

(MR) 3   **a** Explain why the area of this rectangle can be written as $(p + q)(r + s)$.

   b Use the diagram to explain why the area of the rectangle can be written as $pr + ps + qr + qs$.

4 Expand the brackets.

   a $(a + b)(c + d)$ **b** $(a + b)(c - d)$ **c** $(a - b)(c + d)$ **d** $(a - b)(c - d)$

(MR) 5   **a** Explain why the area of the blue rectangle can be written as $(w - x)(y + z)$.

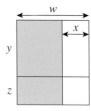

   b Use the diagram to explain why the area of the blue rectangle also can be written as $wy + wz - xy - xz$.

6 Expand the brackets.

   a $x(x + 2)$ **b** $a(a - 3)$ **c** $p(p + 5)$ **d** $y(y - 5)$

7 Expand the brackets. Simplify the result by collecting like terms.

   a $(a + 2)(a + 1)$ **b** $(n + 4)(n + 3)$ **c** $(r + 1)(r + 6)$ **d** $(x + 5)(x - 2)$

   e $(y + 3)(y - 4)$ **f** $(m - 4)(m + 6)$ **g** $(x - 3)(x - 4)$ **h** $(z - 10)(z - 4)$

(MR) 8 Use this diagram to explain why $(x + 3)(x + 5)$ is equivalent to $x^2 + 8x + 15$.

(MR) **9** Use this diagram to explain why $(x + 6)(x - 2)$ is equivalent to $x^2 + 4x - 12$.

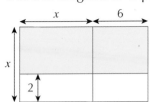

10 Expand the brackets. Simplify each answer as much as possible.

   a $(x + 3)(2 + x)$ **b** $(4 + t)(5 + t)$ **c** $(6 + x)(x + 1)$ **d** $(3 + r)(r - 7)$

   e $(x - 5)(2 + x)$ **f** $(4 + t)(2 - t)$ **g** $(3 - n)(5 + n)$ **h** $(8 - y)(2 - y)$

(PS) **11** A number is missing from each expansion.

Work out the missing number in each case.

   a $(x + 6)(x + 3) = x^2 + \ldots x + 18$ **b** $(x + 3)(x + 2) = x^2 + \ldots x + 6$

   c $(x + 2)(x - 5) = x^2 - \ldots x - 10$ **d** $(x - 3)(x - 1) = x^2 - \ldots x + 3$

12 You can write $(x + 2)(x + 2)$ as $(x + 2)^2$.

Expand these brackets and simplify as much as possible.

   a $(x + 2)^2$ **b** $(x + 1)^2$ **c** $(x + 4)^2$ **d** $(x + 7)^2$

Investigation: The difference of two squares

A Copy and complete each calculation.

   a $5^2 - 4^2 = \ldots$ **b** $12^2 - 11^2 = \ldots$ **c** $20^2 - 19^2 = \ldots$ **d** $35^2 - 34^2 = \ldots$

B Describe any pattern you can see, in your answers to Question **A**.

C Without using a calculator, work out $1000^2 - 999^2$.

D $(n + 1)^2$ is the same as $(n + 1)(n + 1)$. Expand and simplify $(n + 1)^2$.

E Use your answer to Question **D** to simplify $(n + 1)^2 - n^2$.

F Use your answer to Question **E** to explain the result you found in Question **B**.

Ready to progress?

I can expand a bracket when there is a variable inside and outside the bracket.
I can factorise an expression when a variable is a factor.

I can expand the product of two brackets.

Review questions

1 Expand the brackets.

   a $3(x - 5)$ **b** $\frac{1}{2}(4a + 12)$ **c** $9(10 - 2x)$ **d** $2.5(6 + k)$

2 Expand the brackets.

   a $x(x + 1)$ **b** $y(2y + 1)$ **c** $z(2z - 3)$ **d** $w(4 - w)$

(PS) **3 a** Show that the area of this triangle is $x(x + 5)$.

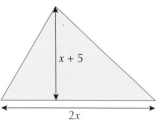

   b Expand the expression $x(x + 5)$.

   c Calculate the area of the triangle when $x = 6$.

4 Multiply out the brackets.

   a $x(x + y)$ **b** $t(2t - 3)$ **c** $3n(n + 6)$ **d** $k(3 - 5k)$

(MR) **5 a** Write down an expression for the area of this rectangle.

$x + 4$ cm

x cm

   b Expand the brackets in your answer to part **a**.

   c Draw a diagram to explain how the area can be divided into two parts to illustrate your answer to part **b**.

6 Factorise each expression as much as possible.

    a $6t - 18$ **b** $t^2 + 4t$ **c** $2x^2 + 7x$ **d** $5x - 3x^2$

    e $3y^2 - 9y$ **f** $8n - 2n^2$ **g** $a^2 + ab$ **h** $x^2 - 2wx$

7 Multiply out the brackets.

    a $2x(2x + 3)$ **b** $3t(4t - 5)$ **c** $2n(3 + 2n)$ **d** $4k(4k - 1)$

8 Multiply out the brackets. Write the results as simply as possible.

    a $(x + 3)(x + 4)$ **b** $(x - 3)(x + 4)$ **c** $(x + 3)(x - 4)$ **d** $(x - 3)(x - 4)$

9 Multiply out the brackets. Write the results as simply as possible.

    a $(n + 3)(5 + n)$ **b** $(d - 3)(2 + d)$ **c** $(1 + t)(10 - t)$ **d** $(x - 8)(x + 9)$

    e $(x + 5)(x - 4)$ **f** $(z - 3)(z - 5)$ **g** $(5 - x)(3 - x)$ **h** $(x + 4)(4 - x)$

10 Expand the brackets in each expression. Write the results as simply as possible.

    a $(x + 3)^2$ **b** $(x - 3)^2$ **c** $(x + 3)(x - 3)$

 11 Look at this expression.

    $(x + 1)^2 + (x - 1)^2$

Expand the brackets and then simplify the answer as much as possible.

 12 a Write down an expression for the area of each rectangle, in square centimetres (cm²).

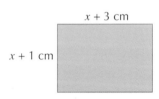
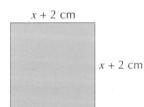

x + 3 cm

x + 1 cm

x + 2 cm

x + 2 cm

    b Show that the difference between the areas of the rectangles is 1 cm².

Challenge
Graphs from expressions

1 The perimeter of this rectangle is 24 cm.
The length of one side is x cm.

   a Write down an expression for the length
of the other side of the rectangle.

   b Explain why the area of the rectangle is $x(12 - x)$ cm^2.

   c Calculate the value of $x(12 - x)$ when:

      i $x = 3$ **ii** $x = 7$.

   d Copy and complete this table. It shows the area for different values of x.

Length of side, x (cm)	1	2	3	4	5	6	7	8	9	10	11
Area, $x(12 - x)$ (cm^2)											

   e Draw a graph to show your
values. Use axes like this.
Two points have been plotted.

   f The points are not in a straight
line. Join the points with a
smooth curve.

   g What is the largest possible
area for the rectangle?

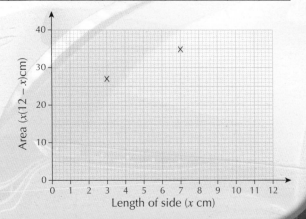

2 Sam fires a rocket vertically up into the air.
The height, in metres, of the rocket t seconds later is given by the expression $30t - 5t^2$.

a Show that the expression can be factorised as $5t(6 - t)$.

b Show that the height after 2 seconds is 40 metres.

c Copy and complete this table.

Time, t (seconds)	1	2	3	4	5	6
Height $5t(6 - t)$ (metres)		40				

d Draw a graph to show the values in your table.
The axes should be similar to those you used in the previous section, with time along the bottom and height up the side. Join the points with a smooth curve.

e What happens after 6 seconds?

f What is the greatest height of the rocket?

3 As a car is travelling, it passes two markers, A and B, 100 metres apart.
As it passes marker A, it is travelling at 18 mph. Its speed gradually increases.

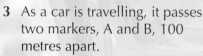

A   B

The distance travelled t seconds after it passes marker A is $t^2 + 5t$ metres.

a Factorise the expression $t^2 + 5t$.

b Show that, three seconds after passing marker A, the car has travelled 24 metres.

c Copy and complete this table showing the distance from A at different times.

Time, t (seconds)	1	2	3	4	5	6	7	8
Distance, $t^2 + 5t$ (metres)			24					

d Draw a graph to show the values in your table.
Use a scale of 1 cm to 1 second on the horizontal axis and 1 cm to 10 metres on the vertical axis.
Join the points with a smooth curve.

e How long does the car take to travel from A to B?

f How can you tell from the graph that the speed of the car is increasing all the time?

9

Decimal numbers

This chapter is going to show you:

- how to extend your ability to work with powers of 10
- when to make suitable rounding and how to use rounded numbers to estimate the results of calculations
- how to use your calculator efficiently.

You should already know:

- how to multiply and divide by 10, 100, 0.1 and 0.01.

About this chapter

One of the earliest aids to arithmetic was the abacus. To use it, you move beads to represent numbers. There are many different types. The Chinese abacus has a number of rods, each holding seven beads. A bar across the rods separates the beads into two sections, with two on one side of the bar and five on the other. The five beads each represent one unit, each of the separate two beads represents five units. So the number seven is represented by one 'five-unit' bead and two 'single-unit' beads.

The Japanese abacus is different, with only four 'single-unit' beads and one 'five-unit' bead on each rod.

Skilful abacus users can calculate very quickly and accurately. There are resources on the internet to show you how to do it.

More recently, though, computers and calculators have been developed to help you work out quite complicated calculations. Modern calculators can do so much more than simple arithmetic. In this chapter you will learn more about the decimal system of counting and practise your skills in using a calculator.

9.1 Powers of 10

Learning objective

Key word

negative power

* To understand and work with both positive and negative powers of ten

You have learnt about positive powers of 10 before. This section will remind you how to use them to solve problems. It will also introduce you to **negative powers** of ten.

This table shows you some powers of 10. Look at it carefully and see if you can spot any patterns.

Power	10^4	10^3	10^2	10^1	10^0	10^{-1}	10^{-2}	10^{-3}	10^{-4}
Value	10 000	1000	100	10	1	0.1	0.01	0.001	0.0001

You should have seen how the negative powers of ten move you into the decimal numbers 0.1, 0.01 and so on.

Note that:

* multiplying by 10^{-1} is the same as multiplying by 0.1, which is the same as dividing by 10

* multiplying by 10^{-2} is the same as multiplying by 0.01, which is the same as dividing by 100

* multiplying by 10^{-3} is the same as dividing by 1000, and so on.

The number next to the negative sign (–) in the power reminds you how many places the digits have to move when you multiply.

* Multiply by a positive power of ten → number gets bigger, digits move to the left.

* Multiply by a negative power of ten → number gets smaller, digits move to the right.

Example 1

Calculate the value of each expression.

a 0.752×10^2 **b** 1.508×10^4 **c** 0.0371×10^6

  a $0.752 \times 10^2 = 0.752 \times 100 = 75.2$

  b $1.508 \times 10^4 = 1.508 \times 10\ 000 = 15\ 080$

  c $0.0371 \times 10^6 = 0.0371 \times 1\ 000\ 000 = 37\ 100$

Example 2

Calculate the value of each expression.

a 3.45×10^{-1} **b** 0.089×10^{-2} **c** 7632×10^{-4}

  a $3.45 \times 10^{-1} = 3.45 \div 10 = 0.345$

  b $0.089 \times 10^{-2} = 0.089 \div 100 = 0.000\ 89$

  c $7632 \times 10^{-4} = 7632 \div 10\ 000 = 0.7632$

Summary

When you are multiplying by 10^n (positive n) you move all the digits n places to the left.
(The decimal point seems to move n places to the right.)
When you are dividing by 10^n (positive n) you move all the digits n places to the right.
(The decimal point seems to move n places to the left.)
When you are multiplying by 10^n (negative n) you move all the digits n places to the right.
(The decimal point seems to move n places to the left.)
When you are dividing by 10^n (negative n) you move all the digits n places to the left.
(The decimal point seems to move n places to the right.)

Exercise 9A

1 Calculate each of these.

   a 6.34×100 **b** $47.3 \div 100$ **c** 66×1000

   d $2.7 \div 1000$ **e** $3076 \times 10\,000$ **f** $7193 \div 10\,000$

   g 9.2×0.1 **h** 0.64×0.01 **i** 0.84×0.001

2 Multiply each number by 10^3.

   a 8.7 **b** 0.32 **c** 103.5 **d** 0.09 **e** 23.06

3 Multiply each number by 10^5.

   a 9.27 **b** 0.82 **c** 706.5 **d** 0.0042 **e** 37.01

4 Multiply each number by 10^{-1}.

   a 7.7 **b** 0.52 **c** 603.5 **d** 7.09 **e** 123.7

5 Multiply each number by 10^{-3}.

   a 8.6 **b** 0.73 **c** 107.4 **d** 3.07 **e** 25.4

6 Multiply each number by 0.01.

   a 2.7 **b** 0.45 **c** 207 **d** 0.08 **e** 41.7

7 Multiply each number by 0.001.

   a 8.7 **b** 0.65 **c** 507 **d** 0.09 **e** 31.7

8 Calculate each of these.

   a 3.76×10^2 **b** $2.3 \div 10^3$ **c** 0.09×10^5

   d $3.09 \div 10^3$ **e** 2.35×10^2 **f** $0.01 \div 10^4$

   g 14.36×10^{-1} **h** 245.3×10^{-3} **i** 1.368×10^{-2}

9 **a** Write down each number as a decimal.

   i 10^{-1} **ii** 10^{-2} **iii** 10^{-3} **iv** 10^{-4}

   b Work each of these out.

   i 9.2×10^{-1} **ii** 0.71×10^{-3} **iii** 45.6×10^{-2}

   iv 4.2×10^{-1} **v** 0.98×10^{-2} **vi** 2.14×10^{-3}

10 The population of Aruba is approximately one hundred thousand. The land area of Aruba is 180 km². What is the average land area per person in Aruba?

  11 The population of India is just over 1.2 billion. The land area of India is 2 973 520 km². What is the average land area per person in India, correct to two significant figures?

  12 The population of Canada is approximately 35 158 000 people. Canada covers an area of approximately ten million square kilometres. What is the average number of people per square kilometre in Canada? Give your answer correct to two significant figures.

 13 China covers an area of approximately ten million square kilometres and has a population of 1.36 billion. What is the average number of people per square kilometre in China?

Activity: Prefixes

You have already met three prefixes that you can use to make decimal multiples of units. They are:

* kilo-, as in kilogram (1000 grams)
* centi-, as in centilitre (one hundredth of a litre)
* milli-, as in millimetre (one thousandth of a metre).

This table gives the main prefixes and their equivalent multiples, written as powers of 10.

Prefix	giga	mega	kilo	centi	milli	micro	nano	pico
Power	10^9	10^6	10^3	10^{-2}	10^{-3}	10^{-6}	10^{-9}	10^{-12}

For example, 7 000 000 000 grams could be written as 7 gigagrams, or – which is more likely – as 7 kilotonnes.

A Use suitable prefixes to write each quantity in a simpler form.

    a 0.004 grams   **b** 8 000 000 watts   **c** 0.75 litres

B Use the internet or a reference book to find out the common abbreviations for these units. For example, kilo is abbreviated to k, as in 6 kg.

C Use the internet or a reference book to find out how far light travels in 1 nanosecond.

9.2 Standard form

Learning objective

• To understand and work with standard form, using both positive and negative powers of ten

You have already used **standard form** to express large numbers.

You know that you can use standard form to write very large numbers, such as those that occur in IT and astronomy, concisely. For example, you can write 73 000 000 000 000 as 7.3×10^{13}.

There are two things to remember about numbers expressed in standard form.

Rule 1: The first part is always a number that is greater than or equal to 1 but less than 10.

Rule 2: The second part is always a power of 10.

$$7.3 \times 10^{13}$$

Examples of numbers that can be written in standard form are:

• the mean radius of the Earth, 6.4×10^6 m

• the speed of light, 2.998×10^8 m/s.

Scientific calculators will display numbers in standard form if the result is too large for the display.

For example, the product of 6 000 000 and 7 000 000 is 42 000 000 000 000 = 4.2×10^{13} could be displayed on your calculator as:

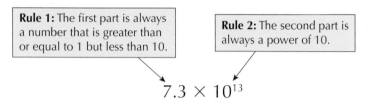

Check to see how your calculator displays this answer.

The idea of standard form can be extended to include negative powers of ten. These are used for very small numbers, such as 0.000 000 67 which can be expressed as 6.7×10^{-7} in standard form.

Note that the power tells you how many places the 6 and 7 need to move, and in which direction, in order to become 6.7.

In the power −7:

• 7 is the number of places the 6 and 7 have to move, to become 6.7

• the negative sign shows that they move to the left.

These two calculators show the same number, but in different ways. Can you write down what the number is?

Example 3

Express each number in standard form.

a 0.076   **b** 0.005 42   **c** 0.000 058 7

a Count how many places the digits have to move so that the most significant digit is between 1 and 10.

In 0.076 that will be two places.

$$0.076 = 7.6 \times 10^{-2}$$

The sign is negative, because they move left.

b Count how many places the digits have to move so that the most significant digit is between 1 and 10.

In 0.005 42 that will be three places.

$$0.005\ 42 = 5.42 \times 10^{-3}$$

The sign is negative, because they move left.

c Count how many places the digits have to move so that the most significant digit is between 1 and 10.

In 0.000 058 7 that will be five places.

$$0.000\ 058\ 7 = 5.87 \times 10^{-5}$$

The sign is negative, because they move left.

When you change a number from standard form, follow the reverse procedure.

Example 4

Write each standard-form number as an ordinary number.

a 5.43×10^{-4} **b** 9.8×10^{-3} **c** 4.521×10^{-6}

a The sign of the power is negative, so when it is written in full, the number will be smaller than 5.43.

You need to move the digits to the right.

Move them through the same number of places as the power of 10, but move them right because of the negative sign.

$$5.43 \times 10^{-4} = 5.43 \div 10\ 000 = 0.000\ 543$$

b Move the digits through the same number of places as the power of 10, but move them right because of the negative sign.

$$9.8 \times 10^{-3} = 98 \div 1000 = 0.0098$$

c Since the power is −6, there will be six zeros before the 4 in the ordinary number.

$$4.521 \times 10^{-6} = 0.000\ 004\ 521$$

 Hint Notice that when changing a standard form number with negative powers, the resulting number will have the same number of zeros before the most significant figure – but you need to include the zero before the decimal point!

Exercise 9B

1 Write each number in standard form.

a 0.85	**b** 0.0127	**c** 0.432	**d** 0.005 12
e 0.95	**f** 0.000 005 5	**g** 0.6	**h** 0.0099
i 0.081	**j** 0.000 000 76	**k** 0.004 321	**l** 0.101 05
m 65 000	**n** 9 897 000	**o** 0.0032	**p** 579 000

2 Write each standard-form number as an ordinary number.

a 6.41×10^{-3}	**b** 9.03×10^{-4}	**c** 8.0×10^{-2}	**d** 7.1×10^{-4}
e 3.142×10^{-3}	**f** 4.5×10^{-2}	**g** 5.01×10^{-4}	**h** 9.852×10^{-7}
i 3.99×10^{-5}	**j** 2.201×10^{-9}	**k** 3.7×10^{-8}	**l** 4.06×10^{-7}
m 5.67×10^{4}	**n** 1.6×10^{6}	**o** 9.64×10^{-5}	**p** 3.07×10^{7}

3 Find the square of each number, giving your answer in standard form.

a 0.08	**b** 0.0015	**c** 0.0012	**d** 0.000 04
e 0.13	**f** 0.000 14	**g** 0.011	**h** 0.0005

4 Write each answer in standard form, correct to two significant figures.

a $3.45 \times 10^{7} + 2.7 \times 10^{7}$ **b** $9.25 \times 10^{-1} + 1.7 \times 10^{-1}$ **c** $2.26 \times 10^{4} + 4.6 \times 10^{5}$

d $1.88 \times 10^{-2} + 5.4 \times 10^{-3}$ **e** $6.49 \times 10^{4} - 2.7 \times 10^{3}$ **f** $3.45 \times 10^{-2} - 5.7 \times 10^{-3}$

g $9.74 \times 10^{7} - 2.1 \times 10^{5}$ **h** $1.43 \times 10^{-5} - 4.9 \times 10^{-6}$ **i** $7.11 \times 10^{-6} - 2.7 \times 10^{-7}$

(PS) 5 The mass of one electron is 0.000 000 000 000 000 000 000 000 000 92 grams.

  a Write this number in standard form.

  b What is the mass of 8 electrons?

  c What is the mass of three million electrons?

6 Explain why, when you multiply a number by 10^{10} you move the digits ten places to the left, but when you multiply by 10^{-10} you move the digits ten places to the right.

(PS) 7 In biology, the diameters of cells are usually measured in microns. (1 micron = 10^{-3} mm)

Yeast cells are little balls about 2 microns in diameter.

Most human cells are about 20 microns in diameter.

  a How long would a string of one million yeast cells be?

  b How long would one thousand human cells be, if they were all in one long string?

(PS) 8 A unit of length used by scientists is the angstrom. (1 angstrom = 10^{-10} m)

A hydrogen atom is about 1 A in diameter, a carbon atom is about 2 A in diameter.

  a What would be the length of a million hydrogen atoms, laid in a row?

  b What would be the length of a string of a billion carbon atoms?

9 A dust mite is 1.25×10^{-2} cm wide and long.

   a What is the width of four dust mites side by side?

   b What is the width of 400 dust mites side by side?

(PS)   **c** A mat measuring 50 cm by 80 cm is full of dust mites. How many dust mites will there be on the mat?

Activity: Division strings

A Calculate each number as accurately as you can. Use standard form where necessary.

$1 \div 11 = 0.090\ 909\ 090\ 909\ 090\ 909 \ldots$

$1 \div 111 =$

$1 \div 1111 =$

$1 \div 11\ 111 =$

$1 \div 111\ 111 =$

b Explain how the pattern continues.

c Use the pattern to write the answer to $1 \div 11\ 111\ 111\ 111\ 111\ 111\ 111$.

9.3 Rounding appropriately

Learning objective

- To round numbers, where necessary, to an appropriate or suitable degree of accuracy

Key word

appropriate degree of accuracy

There are two main reasons for rounding, both of which you have met before.

- One reason is to give an answer to an **appropriate degree of accuracy**.
- The other reason is to enable you to make an estimate of the answer to a problem.

Example 5

Round each number to:

i one decimal place   **ii** two decimal places.

   a 7.356   **b** 13.978   **c** 0.2387

   a i 7.356 = 7.4 to one decimal place   **ii** 7.356 = 7.36 to two decimal places
   b i 13.978 = 14.0 to one decimal place   **ii** 13.978 = 13.98 to two decimal places
   c i 0.2387 = 0.2 to one decimal place   **ii** 0.2387 = 0.24 to two decimal places

---

Example 6

Round each number to one significant figure.

a 18.67   **b** 0.037 61   **c** 7.95

a 18.67 = 20 to one significant figure

b 0.037 61 = 0.04 to one significant figure

c 7.95 = 8 to one significant figure

Example 7

Estimate the answer to each calculation.

a 21% of £598   **b** $\dfrac{23.7 + 69.3}{3.1 \times 5.2}$ **c** $3.9^2 \div 0.0378$

The method is to round the numbers to one significant figure each time.

a 21% of £598 ≈ 20% of £600 = £120

b $\dfrac{23.7 + 69.3}{3.1 \times 5.2} \approx \dfrac{20 + 70}{3 \times 5} = \dfrac{90}{15} = 6$

c $3.9^2 \div 0.0378 \approx 4^2 \div 0.04 = 16 \div 0.04 = 160 \div 0.4 = 1600 \div 4 = 400$

Example 8

Which measurement has been rounded to the most appropriate degree of accuracy?

a The distance from my house to the local post office

     i 721.4 m   **ii** 721 m or   **iii** 700 m

b The time in which an athlete runs a 100 m race

     i 10.14 seconds   **ii** 10.1 seconds   **iii** 10 seconds

c 45 ÷ 37 =

     i 1   **ii** 1.22   **iii** 1.216 216 2

a Distances are usually rounded to one or two significant figures.

Hence, 700 m is the most appropriate answer.

b 100-metre times need to be accurate to the nearest hundredth of a second.

Hence, 10.14 s is the most appropriate answer.

c An answer to a calculation is usually given to either the same accuracy as the numbers in the calculation itself or just one more significant figure.

The context sometimes helps you decide. Here there is no context to help, so as the figures in the calculation are correct to two significant figures, it is appropriate to round to three significant figures.

Hence 1.22 is the most suitable answer.

Exercise 9C

1 Round each number to:

  i one decimal place   **ii** two decimal places.

    a 2.367   **b** 13.0813   **c** 8.907   **d** 20.029

    e 0.999   **f** 4.0599   **g** 0.853   **h** 3.141 59

2 Round each number to one significant figure.

    a 4560   **b** 0.0941   **c** 3.098   **d** 42.611

    e 0.999   **f** 505.98   **g** 34.72   **h** 3.141 59

3 Write each number to two significant figures.

    a 19.6%   **b** £273   **c** 2.83   **d** 0.0185

    e 12.73   **f** 0.058 61   **g** 23.19   **h** 573

4 Estimate the answer to each calculation.

    a 19% of £278   **b** $23.2 \div 0.018$ **c** $12.3^2 \times 0.058$ **d** $\dfrac{23.1 + 57.3}{16.5 + 7.3}$

5 Estimate the answer to each calculation.

  Where appropriate, round the answer to a suitable degree of accuracy.

    a $\dfrac{0.245 \times 0.03}{1.89 \times 3.14}$ **b** $\dfrac{45.9 \times 83.2}{26.7 - 9.8}$ **c** 14% of 450 kg

    d $59.5 \div 0.13^2$ **e** $(3.95 \times 0.68)^2$ **f** 28% of 621 km

    g 4% of £812   **h** 0.068×0.032 **i** $1.86 \times 10^{-5} + 5.14 \times 10^{-5}$

6 Round each quantity to a sensible degree of accuracy.

    a Average speed of a journey: 63.7 mph

    b Size of an angle in a right-angled triangle: 23.478°

    c Mass of a sack of potatoes: 46.89 kg

    d Time taken to boil an egg: 4 minutes 3.7 seconds

    e Time to run a marathon: 2 hours 32 minutes and 44 seconds

    f World record for running 100 m: 9.58 seconds

7 Use a calculator to work out each answer, then round each result to an appropriate degree of accuracy.

    a $\dfrac{56.2 + 48.9}{17.8 - 12.5}$ **b** $\dfrac{12.7 \times 13.9}{8.9 \times 4.3}$ **c** $1 \div 32$

    d 0.58^2 **e** $1 \div 45$ **f** $23.478 \div 0.123$

8 Billy rounded a number to 8. His brother Isaac rounded the same number to 7.8 but they were both correct. Explain how this could be.

9 A shop sells rolls of wire with 46 metres of wire on them.

Helen needs to be able to create 9.1 cm lengths of wire for some artwork with her class.

Estimate how many 9.1 cm lengths she could cut from the 46 m roll.

9.4 Mental calculations

Learning objective

- To learn and understand some routines that can help in mental arithmetic

There will be many occasions when you need to do a calculation mentally, as you have no calculator or paper and pencil with you.

You should already know your times tables up to twelve and be able to recall them easily. You should be able to do simple calculations in your head. You can estimate solutions too, but there are some simple routines that will help you.

- $\times 25$ Just divide by 4 then multiply by 100 (or multiply by 100 then divide by 4).
- $\div 25$ Just multiply by 4 then divide by 100 (or divide by 100 then multiply by 4).
- $\div 20$ Just halve the number then divide by 10.
- $\div 16$ Just halve, then halve, then halve, then halve again (four lots of halving).
- $\times 15$ Multiply by 3, halve the answer, then multiply by 10 (or halve, then multiply by 3, then $\times 10$).

There are many methods of mental calculation, based on routines similar to these.

Example 9

Work these out mentally.

a 88×25 **b** $900 \div 25$ **c** 18×15

a 88×25	Divide 88 by 4 to get 22, multiply by 100 to get 2200.	$88 \times 25 = 2200$
b $900 \div 25$	Divide by 100 to get 9, multiply by 4 to get 36.	$900 \div 25 = 36$
c 18×15	Halve to get 9, multiply by 3 to get 27, multiply by 10 to get 270.	$18 \times 15 = 270$

Example 10

Work out 13.6×0.25 without using a calculator.

First, think of the problem as 136×25 (and remember there are three decimal places).

Now divide 136 by 4.   Do this by halving and halving again.

136 halved is 68, halved again is 34.

Now multiply 34 by 100 to get 3400.

Now restore the three decimal places, to give 3.4 as the answer.

Hence $13.6 \times 0.25 = 3.4$.

Example 11

Work these out mentally.

a $8 \times £1.99$ **b** $16 \times £2.98$ **c** $18 \times £4.95$

a $8 \times £1.99$	Treat as $8 \times (£2 - 1p) = £16 - 8p = £15.92$	$8 \times £1.99 = £15.92$
b $16 \times £2.98$	Treat as $16 \times (£3 - 2p) = £48 - 32p = £47.68$	$16 \times £2.98 = £47.68$
c $18 \times £4.95$	Treat as $18 \times (£5 - 5p) = £90 - 90p = £89.10$	$18 \times £4.95 = £89.10$

Exercise 9D

Do not use a calculator to answer any of these questions.

1 Work out each of these in your head. Then write down the answer.

   a 400×25 **b** 72×25 **c** 108×25 **d** 8.4×25

   e $600 \div 25$ **f** $450 \div 25$ **g** $1200 \div 25$ **h** $1 \div 25$

   i 28×15 **j** 82×15 **k** 104×15 **l** 36×15

   m $640 \div 20$ **n** $700 \div 20$ **o** $800 \div 16$ **p** $2000 \div 16$

2 Work out each of these in your head. Then write down the answer.

   a $900 \div 20$ **b** 52×25 **c** 110×15 **d** $1200 \div 16$

   e $700 \div 25$ **f** $350 \div 20$ **g** 36×25 **h** 68×15

   i $2800 \div 16$ **j** $900 \div 25$ **k** $340 \div 20$ **l** 76×25

   m 2.4×15 **n** $848 \div 16$ **o** $7000 \div 25$ **p** $200 \div 16$

3 Work out the answer to each of these. Use the method you like best.

   a $72 \div 20$ **b** 1.8×1.5 **c** 2.8×2.5 **d** $1.24 \div 2.5$

   e $0.8 \div 0.16$ **f** 12.6×0.15 **g** $2.6 \div 2.5$ **h** 9.6×0.25

4 By suitable rounding, estimate each of these in your head. Then write down the answer.

   a $678 \div 19$ **b** 92×24 **c** 310×14.5 **d** $219 \div 16.3$

   e $452 \div 25.2$ **f** $750 \div 18.3$ **g** 46×23.7 **h** 78×13.9

   i $2370 \div 15.3$ **j** $908 \div 25.7$ **k** $530 \div 19.3$ **l** 7.6×2.6

   m 7.4×1.45 **n** $95.6 \div 1.56$ **o** $7.3 \div 0.24$ **p** $2.6 \div 0.17$

5 Work out each of these in your head. Then write down the answer.

    a $12 \times £1.99$ **b** $15 \times £2.98$ **c** $18 \times £1.95$ **d** $8 \times £3.99$

    e $16 \times £4.98$ **f** $7 \times £6.95$ **g** $8 \times £7.99$ **h** $27 \times £9.98$

6 Work these answers out in your head and just write down the answers.

    a A bar of chocolate costs one pound and eighty pence. How much would twenty-five bars cost?

    b Subtract one hundred and seventeen from one hundred and seventy-two.

    c What is the cost of thirteen books at £2.99 each?

    d A school has twice as many girls as boys. There are 286 girls. How many pupils are there in the school?

    e A lottery win of £8200 is shared among twenty-five people. How much does each person get?

(PS) 7 A rectangle measures 2.46 m by 0.27 m.

What is the approximate area of the rectangle?

(PS) 8 Approximately how much will it cost to pay for a party of 215 guests at a charge of £8.95 each?

(PS) 9 I buy 42.8 litres of petrol at £1.29 per litre.

What is the approximate cost?

(PS) 10 The *Daily Mail* reported that in 2013 it sold an average of 1 863 000 copies per working day.

Approximately, how many copies did it sell in the year?

Investigation: Subtracting squares

A Work these out.

    a 1.2^2 **b** 1.3^2 **c** $1.3^2 - 1.2^2$ **d** 0.1×2.5

B Work these out.

    a 1.1^2 **b** 1.5^2 **c** $1.5^2 - 1.1^2$ **d** 0.4×2.6

C Look for a connection between the calculations in parts **c** and **d** of questions **A** and **B**.

Then write down the answer to $3.1^2 - 2.1^2$. Check your answer with a calculator.

9.5 Solving problems

Learning objective

- To solve real-life problems involving multiplication or division

You will use decimal calculations on many occasions in everyday life, for example, when you are shopping. People at work often need to do some calculations while solving business problems.

Work through the next two examples to use what you have learned so far, when you are solving some problems.

Example 12

Which jar of jam offers the better value?

The smaller jar gives $454 \div 89 = 5.1$ grams/penny.

The larger jar gives $2000 \div 400 = 5$ grams/penny.

So, the smaller jar offers the better value.

Example 13

The mass of a bag of identical marbles is 375 g.

Susie takes out seven marbles.

The mass of the bag is then 270 g.

How many marbles were there in the bag to start with?

The difference in mass is $375 - 270 = 105$ g.

This is the mass of seven marbles, so the mass of one marble is:

$105 \div 7 = 15$ g

Hence, the number of marbles originally in the bag is given by:

$375 \div 15 = 25$

Exercise 9E

  1 A school buys some books as prizes. The books cost £4.25 each.

The school had £40 to spend on prizes. They buy as many books as possible.

How much money is left?

  2 Sticky labels cost £33.40 per box.

In the box there are 150 sheets. Each sheet has 14 labels on it.

What is the cost of one label, correct to the nearest penny?

MR **3**
FS

Two families went to the cinema.

It cost the Ahmed family of one adult and two children £11.50.

It cost the Smith family of two adults and two children £16.

What is the cost of an adult's ticket and a child's ticket?

MR **4**
PS

This is the charge for hiring a car.

a How much will it cost to hire the car for five days?

b John pays £137.00 for car hire. How many days did he have the car?

CAR HIRE
£25 PLUS
£16.00 PER DAY

PS **5**

Leisureways sell Kayenno trainers for £69.99 but give 10% discount for Jiggers Running Club members.

All Sports sell the same trainers for £75 but are having a sale in which everything is reduced by 15%. In which shop are the trainers cheaper for Jiggers Running Club members?

PS **6**

A supermarket sells crisps in different-sized packets.

An ordinary bag contains 30 g and costs 28p.

A large bag contains 100 g and costs 90p.

A jumbo bag contains 250 g and costs £2.30.

Which bag is the best value? You must show all your working.

PS **7**

A cash-and-carry sells crisps in boxes.

A 12-packet box costs £3.00.

An 18-packet box costs £5.00.

A 30-packet box costs £8.00.

Which box gives the best value?

8

A box of chocolates has three soft-centred chocolates for every two hard-centred chocolates.

There are 40 chocolates altogether in the box. How many of them are soft-centred?

9

Davy is twice as old as Arnie. The sum of their ages is 36 years.

How old are they?

PS **10**

A recipe for marmalade uses 65 g of oranges for every 100 g of marmalade. Mary has 10 kg of oranges.
How many 454-gram jars of marmalade can Mary make?

11 Fibonacci sequences are formed by adding the previous two numbers to get the next number. For example:

1, 1, 2, 3, 5, 8, 13, ...

a Write down the next three terms of the Fibonacci sequence that starts:

2, 2, 4, 6, 10, 16, ...

b Work out the missing values of this sequence:

2, 3, 5, ..., 13, ..., ..., 55, ...

c Work out the first three terms of this sequence:

..., ..., ..., 0, 1, 1, 2, 3, 5, 8, ...

12 What fraction is halfway between $\frac{3}{5}$ and $\frac{9}{10}$?

Investigation: Two-digit numbers

A Choose a two-digit number such as 18.

Multiply the units digit by 3 and add the tens digit, which give:

$3 \times 8 + 1 = 25$

B Repeat with the new number:

$3 \times 5 + 2 = 17$

Keep repeating the procedure until the numbers start repeating, namely:

$17 \rightarrow 22 \rightarrow 8 \rightarrow 24 \rightarrow ...$

C Show the chains on a poster. For example:

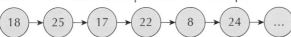

Ready to progress?

I can multiply and divide by simple powers of 10.
I can round numbers correct to one decimal place.

I can round numbers correct to two or more decimal places.
I can multiply and divide by any positive power of 10.
I can calculate problems mentally, using some shortcuts.

I understand standard form numbers with both positive and negative powers of 10.
I can round numbers to one significant figure in order to make sensible estimations.
I can solve best-value problems.

Review questions

1 a Oliver's height is 0.95 m. Aysha is 0.3 m taller than Peter. What is Aysha's height?

  b Rashid's height is 1.35 m. Enya is 0.2 m shorter than Rashid. What is Enya's height?

  c Carla's height is 1.8 m. What is Carla's height in centimetres?

(PS) (FS) 2 A shop sells kitchen rolls.

You can buy them in packs of nine or packs of six.

Which pack gives you better value for money?

Packet of 9 kitchenrolls
£4.80

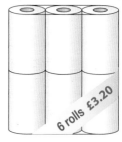

Packet of 6 kitchen rolls
£3.20

(PS) (FS) 3 Two companies design and sell leaflets.

The table shows how much the companies charge.

		Company	
		A	B
Costs (£)	Design leaflet	35.00	45.00
	Print first 500	65.00	45.00
	Print each extra 100	2.00	5.00
	Postage	5.50	4.50

I want 900 leaflets.

a Which company gives me the cheaper deal?

b How much cheaper is it?

4 The value of Euler's number e is 2.718 282, correct to six decimal places.

    a Write the value of e correct to four decimal places.

    b Which value below is closest to the value of e?

$$\frac{100}{37} \qquad\qquad 2\frac{8}{11} \qquad\qquad \left(\frac{8}{5}\right)^2 \qquad\qquad \frac{590}{217}$$

5 This is the method for finding the geometric mean of two numbers.

- Multiply the two numbers together.
- Then find the square root of the result.

For example, the geometric mean of 4 and 16 = $\sqrt{(4 \times 16)} = \sqrt{64} = 8$.

    a Two numbers have a geometric mean of 20. One number is 8. What is the other?

    b Lynn says: 'I don't think that the numbers –4 and 5 have a geometric mean.'
        Explain why Lynn is correct.

6 Copy these multiplication grids and fill in the missing numbers.

×	6	
7	42	
–5		40

×	0.3	
4		2.4
		5.4

7 In a right-angled triangle, the lengths of the two shorter sides are 3.15×10^5 mm and 1.87×10^5 mm.

What is the length of the hypotenuse? Give your answer in standard form and correct to two significant figures.

8 A shop sells porridge oats in packs of different sizes.

Small	700 g	£0.80
Medium	1 kg	£1.10
Large	1.6 kg	£1.80

Which pack represents the best buy?

9 A recipe for flapjack uses 225 g of oats to make 8 wedges of flapjack.

How many oats are needed to make 20 wedges of flapjack?

Mathematical reasoning
Paper

Paper sizes

Standard paper sizes are called A0, A1, A2 and so on.
A sheet of A0 is approximately 1188 mm by 840 mm.
The next size, A1, is made by cutting A0 in half.
This is a diagram of a piece of A0 paper.

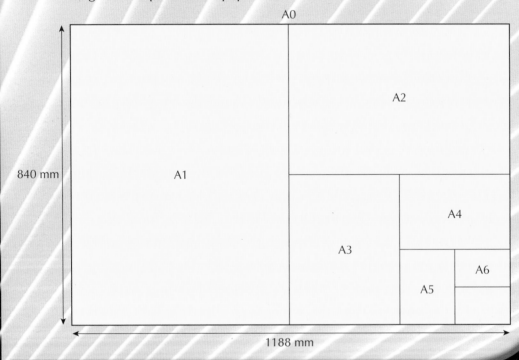

9 Decimal numbers

Paper for sale 1 ream = 500 sheets

Reams of paper – Special offer

Grade of paper	Prices		
	1 ream	5–9 reams	10+ reams
Standard	£2.10	SAVE $\frac{1}{3}$ per ream	EXTRA 10% discount
Special	£1.80		
High-quality	£3.00		
Photo	£3.60		

Example for standard paper

1 ream costs £2.10

5 reams at full price = 5 × £2.10 = £10.50

Saving = $\frac{1}{3}$ of £10.50 = £3.50

Cost of 5 reams = £7.00

10 reams with $\frac{1}{3}$ off = 2 × £7.00 = £14.00

Extra 10% discount = £1.40

Cost of 10 reams = £12.60

> **Buy 5 reams**
> You save £3.50
>
> **Buy 10 reams**
> You save £8.40

Use the information on the opposite page to answer these questions.

1 How many sheets of paper are there in 40 reams of paper?

2 The thickness of a piece of paper is 0.12 mm.
How high would a stack of 4 reams (2000 sheets) of this paper be?
Give your answer in centimetres.

3 This is a table of paper sizes.

Paper size	A0	A1	A2	A3	A4	A5
Length (mm)	1188	840	594			
Width (mm)	840	594				

Copy and complete the table.
Check your result for A4 paper by measuring.

4 How many pieces of A5 paper would you need, to make a piece of A1 paper?

5 Look at the special offers and work out the cost of:

a 3 reams of special grade paper

b 5 reams of high-quality grade paper

c 10 reams of photo grade paper.

6 Work out the saving if you use the offers to buy 20 reams of high-quality paper.

10

Prisms and cylinders

This chapter is going to show you:

- how to convert from one metric unit to another for area and volume
- how to calculate the surface area and the volume of a prism
- how to calculate the surface area and the volume of a cylinder.

You should already know:

- the metric units for area and volume
- how to calculate areas of 2D shapes
- how to calculate the surface area and volume of a cuboid.

About this chapter

Look at the photo carefully. Why is every cell in this honeycomb a hexagon?

Honeycombs store honey. Honey is obviously valuable to bees. It feeds their young. It sustains the hive. It makes the wax that holds the honeycomb together. It takes thousands and thousands of bee hours, tens of thousands of flights across the meadow, to gather nectar from flower after flower after flower, so it's reasonable to suppose that back at the hive, bees want a tight, secure storage structure that is as simple to build as possible. So which to choose? A triangle? A square? Or a hexagon? Which one is best?

Compactness matters. The more compact your structure, the less wax you need to construct the honeycomb. Wax is expensive. A bee must consume about eight ounces of honey to produce a single ounce of wax. So you want the most compact building shape you can find. This is a hexagon!

10.1 Metric units for area and volume

Learning objective

- To convert from one metric unit to another

Key words

conversion	conversion factor
hectare	

You need to know the metric units for area, volume and capacity.

They are listed in the table, which also gives the **conversions** between these units.

Area	Volume	Capacity
$10\ 000\ \text{m}^2 = 1$ **hectare** (ha)	$1\ 000\ 000\ \text{cm}^3 = 1\ \text{m}^3$	$1\ \text{m}^3 = 1000$ litres (l or l)
$10\ 000\ \text{cm}^2 = 1\ \text{m}^2$	$1000\ \text{mm}^3 = 1\ \text{cm}^3$	$1000\ \text{cm}^3 = 1$ litre
$1\ \text{m}^2 = 1000\ 000\ \text{mm}^2$		$1\ \text{cm}^3 = 1$ millilitre (ml or ml)
$100\ \text{mm}^2 = 1\ \text{cm}^2$		10 millilitres $= 1$ centilitre (cl or cl)
		1000 millilitres $= 100$ centilitres $= 1$ litre

The unit symbol for litres is the letter l, which is often written as l, as shown in the table.

To avoid confusion with the digit 1 (one), it is also common to use the full unit name instead of the symbol.

Remember:

- to convert large units to smaller units, always multiply by the **conversion factor**
- to convert small units to larger units, always divide by the conversion factor.

Example 1

Convert each of these units as indicated.

a $72\ 000\ \text{cm}^2$ to m^2 **b** $0.3\ \text{cm}^3$ to mm^3 **c** $4500\ \text{cm}^3$ to litres

a You are converting small units to larger units, so divide by the conversion factor 10 000.
$$72\ 000\ \text{cm}^2 = 72\ 000 \div 10\ 000 = 7.2\ \text{m}^2$$

b You are converting large units to smaller units, so multiply by the conversion factor 1000.
$$0.3\ \text{cm}^3 = 0.3 \times 1000 = 300\ \text{mm}^3$$

c You are converting small units to larger units, so divide by the conversion factor 1000.
$$4500\ \text{cm}^3 = 4500 \div 1000 = 4.5\ \text{litres}$$

Exercise 10A

1 Express each of these in square centimetres (cm^2).

    a $4\ \text{m}^2$ **b** $7\ \text{m}^2$ **c** $20\ \text{m}^2$ **d** $3.5\ \text{m}^2$ **e** $0.8\ \text{m}^2$

2 Express each of these in square millimetres (mm^2).

    a $2\ \text{cm}^2$ **b** $5\ \text{cm}^2$ **c** $8.5\ \text{cm}^2$ **d** $36\ \text{cm}^2$ **e** $0.4\ \text{cm}^2$

3 Express each of these in square centimetres (cm^2).

  a 800 mm^2 **b** 2500 mm^2 **c** 7830 mm^2 **d** 540 mm^2 **e** 60 mm^2

4 Express each of these in square metres (m^2).

  a 20 000 cm^2 **b** 85 000 cm^2 **c** 270 000 cm^2 **d** 18 600 cm^2 **e** 3480 cm^2

5 Express each of these in cubic millimetres (mm^3).

  a 3 cm^3 **b** 10 cm^3 **c** 6.8 cm^3 **d** 0.3 cm^3 **e** 0.48 cm^3

6 Express each of these in cubic metres (m^3).

  a 5 000 000 cm^3 **b** 7 500 000 cm^3 **c** 12 000 000 cm^3
  d 650 000 cm^3 **e** 2000 cm^3

7 Express each of these in litres.

  a 8000 cm^3 **b** 17 000 cm^3 **c** 500 cm^3 **d** 3 m^3 **e** 7.2 m^3

8 Express each measure as indicated.

  a 85 ml in cl   **b** 1.2 litres in cl   **c** 8.4 cl in ml
  d 4500 ml in litres   **e** 2.4 litres in ml

9 How many square paving slabs, each of side 50 cm, are needed to cover a rectangular yard measuring 8 m by 5 m?

10 A football pitch measures 120 m by 90 m.

  Find the area of the pitch in:

  a square metres (m^2)   **b** hectares (ha).

11 A fish tank is 1.5 m long, 40 cm wide and 25 cm high.

  How many litres of water will it hold if it is filled to the top?

(MR) 12 The volume of the cough-medicine bottle is 240 cm^3.

  How many days will the cough medicine last?

(MR) 13 How many lead cubes of side 2 cm can be cast from 4 litres of molten lead?

A farmer has 100 m of fencing to enclose his sheep.

He uses the wall for one side of the rectangular sheep-pen.

If each sheep requires 5 m² of grass inside the pen, what is the greatest number of sheep that the pen can hold?

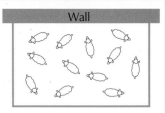

Wall

10.2 Volume of a prism

Learning objective

- To calculate the volume of a prism

A **prism** is a three-dimensional (3D) shape that has exactly the same two-dimensional (2D) shape running all the way through it, whenever it is cut across, perpendicular to its length.

This 2D shape is called the **cross-section** of the prism.

The shape of the cross-section depends on the type of prism, but it is always the same for a particular prism.

You can work out the volume, V, of a prism by multiplying the area, A, of its cross-section by the length, l, of the prism or its height, h, if it stands on one end.

$$V = Al \qquad \text{or} \qquad V = Ah$$

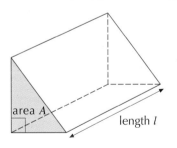

area A length l

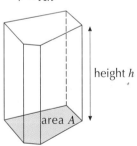

height h

area A

Example 2

Calculate the volume of this triangular prism.

The cross-section is a right-angled triangle with an area of $\frac{6 \times 8}{2} = 24$ cm².

So the volume is given by:

V = area of cross-section × length = $24 \times 15 = 360$ cm³

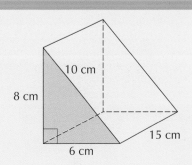

10 cm

8 cm

6 cm

15 cm

Example 3

The area of the cross-section of this hexagonal prism is 25 cm².

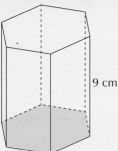

9 cm

Calculate the volume of the prism.

The volume is given by:

V = area of cross-section × height = $25 \times 9 = 225$ cm³

Exercise 10B

1. Calculate the volume of each prism.

a

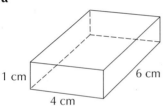

1 cm   4 cm   6 cm

b

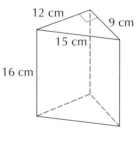

12 cm   9 cm   15 cm   16 cm

c

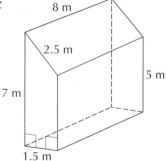

8 m   2.5 m   5 m   7 m   1.5 m

2. The cross-section of the pencil shown is a hexagonal prism with an area of 50 mm².
The length of the pencil is 160 mm.

a Calculate the volume of the pencil, in cubic millimetres.

b Write down the volume of the pencil, in cubic centimetres.

3. The biscuit tin shown is an octagonal prism with a cross-sectional area of 350 cm²
and a height of 9 cm.

Calculate the volume of the tin.

4 The box of a chocolate bar is an equilateral triangular prism.

The area of its cross-section is 15.5 cm^2 and the length of the prism is 30 cm.

Calculate the volume of the box.

5 The diagram shows the cross-section of a swimming pool, along its length. The pool is 15 m wide.

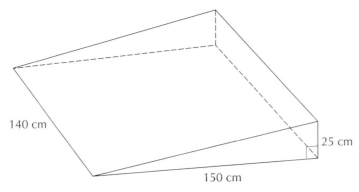

a Calculate the area of cross-section of the pool.

b Work out the volume of the pool.

c How many litres of water does the pool hold when it is full?

(MR) **6** Leroy is making a solid concrete ramp for wheelchair-access to his house.

The dimensions of the ramp are shown on the diagram.

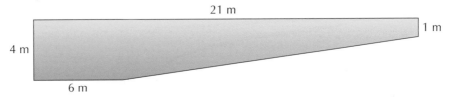

a Calculate the volume of the ramp, giving your answer in cubic centimetres.

b One cubic metre of cement weighs 2.4 tonnes.

What is the mass of concrete that Leroy uses?

(PS) **7** The cross-section of a block of wood is a trapezium.

The height of the trapezium is 12 cm and the volume of the block is 9600 cm^3.

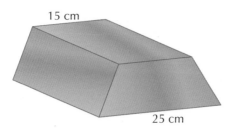

Calculate the length of the block.

A A doll's house is made from a cuboid with an isosceles triangular prism on top.
Calculate the volume of the doll's house.

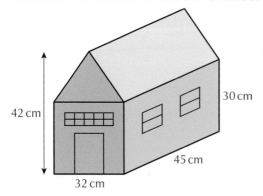

42 cm
30 cm
45 cm
32 cm

B Calculate the volume of this 3D letter-box shape.

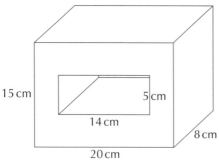

15 cm
5 cm
14 cm
8 cm
20 cm

10.3 Surface area of a prism

Learning objective

• To calculate the surface area of a prism

You can find the **surface area** of a prism by calculating the sum of the areas of its faces.

For example, in the prism shown here, its **total surface area** is made up of the two end pentagons plus five rectangles.

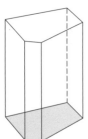

Example 4

Calculate the total surface area of this triangular prism.

The cross-section is a right-angled triangle with an area of $\frac{6 \times 8}{2} = 24$ cm^2.

So the total area of the two end triangles is 48 cm^2.

The sum of the areas of the three rectangles is:

$(6 \times 15) + (8 \times 15) + (10 \times 15) = 360$ cm^2

So, the total surface area is:

$48 + 360 = 408$ cm^2

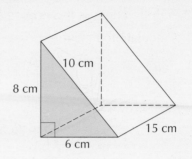

Exercise 10C

1 Calculate the total surface area of each prism.

a
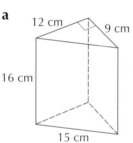
12 cm
9 cm
16 cm
15 cm

b
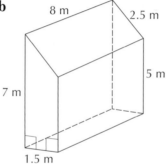
8 m
2.5 m
5 m
7 m
1.5 m

c
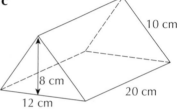
10 cm
8 cm
12 cm
20 cm

2 The cross-section of this prism is a trapezium.

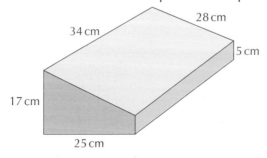
28 cm
34 cm
5 cm
17 cm
25 cm

    a Calculate the area of the trapezium.

    b Calculate the total surface area of the prism.

3 This regular octagonal prism has a cross-sectional area of 96 cm^2 and a length of 32 cm.

Each edge of the octagon is 6 cm long.

Calculate the total surface area of the prism.

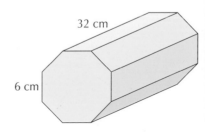
32 cm
6 cm

4 A tent is in the shape of a triangular prism.

Its length is 2.4 m, its height is 1.6 m, the width of the triangular end is 2.4 m and the length of the sloping side of the triangular end is 2 m.

Calculate the surface area of the outside of the tent.

5 This gift box is in the form of an equilateral triangular prism.

The cross-sectional area is 27.7 cm^2.

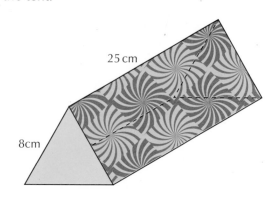

25 cm

8cm

a Calculate the area of the net needed to make the box.

Do not include the area of the tabs.

b Calculate the volume of the box.

 6 The cross-sectional area of the prism shown is made from five equal squares, each with side length of 4 cm.

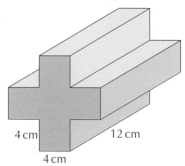

4 cm

12 cm

4 cm

The prism is 12 cm long.

a Calculate the area of the cross-section.

b Calculate the total surface area of the prism.

c Calculate the volume of the prism.

 7 The total surface area of this isosceles triangular prism is 896 cm^2.

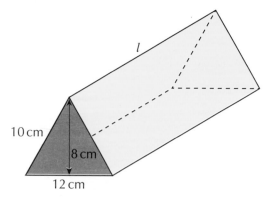

l

10 cm

8 cm

12 cm

Work out the length, *l*, of the prism.

Investigation: Painted cubes

A 2 by 2 by 2 cube is made from eight smaller yellow cubes, as shown.

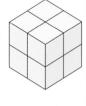

The outside of the large cube is painted red and then the large cube is taken apart.

A **a** How many of the smaller cubes have no faces painted red?

   b How many of the smaller cubes have one face painted red?

   c How many of the smaller cubes have two faces painted red?

   d How many of the smaller cubes have three faces painted red?

B Now a 3 by 3 by 3 cube is made from 27 smaller yellow cubes, as shown.

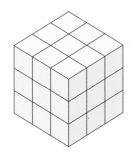

   Again, the outside of the large cube is painted red and then the large cube is taken apart.

   a How many of the smaller cubes have no faces painted red?

   b How many of the smaller cubes have one face painted red?

   c How many of the smaller cubes have two faces painted red?

   d How many of the smaller cubes have three faces painted red?

C Now repeat for a 4 by 4 by 4 cube and a 5 by 5 by 5 cube.

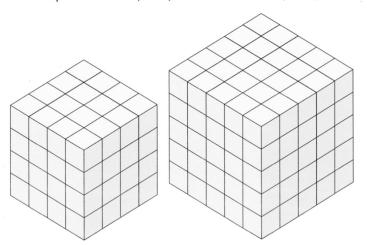

D Copy and complete this table.

Size of yellow cube	No faces painted red	One face painted red	Two faces painted red	Three faces painted red
2 by 2 by 2				
3 by 3 by 3				
4 by 4 by 4				
5 by 5 by 5				

E You now have a 6 by 6 by 6 cube. It is treated exactly as the previous cubes.

   a How many of the smaller cubes have no faces painted red?

   b How many of the smaller cubes have one face painted red?

   c How many of the smaller cubes have two faces painted red?

   d How many of the smaller cubes have three faces painted red?

10.4 Volume of a cylinder

Learning objective

- To calculate the volume of a cylinder

A cylinder is a circular prism.

The cross-section of a cylinder is a circle of radius r.

The area of the cross-section is A.

$A = \pi r^2$

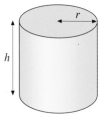

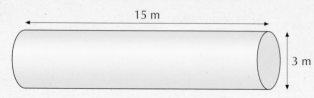

If the height of the cylinder is h,

then the volume, V, of the cylinder is given by:

$$V = \pi r^2 \times h = \pi r^2 h$$

If the length of the cylinder is l,

then the volume, V, of the cylinder is given by:

$$V = \pi r^2 \times l = \pi r^2 l$$

Example 5

Calculate the volume of each cylinder, giving your answers correct to one decimal place.

a
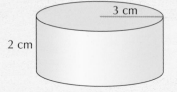
3 cm
2 cm

b
15 m
3 m

a $V = \pi r^2 h = \pi \times 3^2 \times 2 = 56.5 \text{ cm}^3$

b $V = \pi r^2 l = \pi \times 1.5^2 \times 15 = 106.0 \text{ m}^3$

Exercise 10D

In this exercise, take $\pi = 3.14$ or use the $\boxed{\pi}$ key on your calculator.

1 Calculate the volume of each cylinder. Give your answers correct to one decimal place.

a
6 cm
10 cm

b
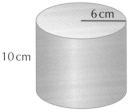
3 cm
8 cm

c

4 m
1.5 m

d
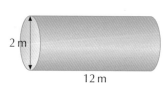
2 m
12 m

e

5 cm
0.5 cm

2 The diameter of a 2p coin is 26 mm and its thickness is 2 mm.

Calculate the volume of the coin.

Give your answer correct to the nearest cubic millimetre.

3 The diagram shows the internal measurements of a cylindrical paddling pool.

a Calculate the volume of the pool.

Give your answer in cubic metres, giving your answer to two decimal places.

b How many litres of water are there in the pool when it is three-quarters full?

Give your answer to the nearest litre.

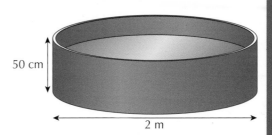

 4 These are three cake tins.

a
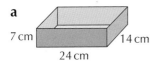

b
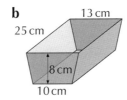

c

Which tin has the greatest volume?

5 The diagram shows a tea urn.

The inner diameter of the cylinder is 30 cm and the inner height is 50 cm.

a Calculate the volume of the inner cylinder.

Give your answer in litres, correct to one decimal place.

b Mr Lloyd is serving tea at a local fête.

A mug holds 20 cl of tea, which leaves space for milk and sugar to be added.

How many mugs of tea can he serve from a full urn?

6 A winners' podium is going to be built at an athletics stadium.

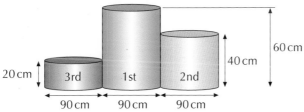

The diagram shows the dimensions of the podium.

a Calculate the volume of each cylinder.

Give your answers correct to the nearest cubic centimetre.

b What is the total volume of the podium?

Give your answer in cubic metres, correct to two decimal places.

7 This paint tin has a capacity of 10 litres.

    a Write down the volume of the tin, in cubic centimetres.

    b Calculate the area of the base of the tin.

    c Calculate the diameter of the base.

      Give your answer correct to one decimal place.

40 cm

Problem solving: Containers

Sophie can buy soup either in cylindrical tins or in cartons.

A soup tin has a diameter of 7 cm and height of 11 cm.

The cartons are cuboid-shaped boxes that are 8 cm long, 5 cm wide and 11 cm high.

A Which container holds more soup?

B Sophie has a cardboard tray in the shape of an open cuboid. It is 56 cm long, 35 cm wide and 2 cm high. What is the maximum number of each container that she could fit in the tray?

C Which container uses the space in the tray more efficiently? Explain how you know.

D What is the best way for a shop to display 15 of each type of container, for a promotion?

10.5 Surface area of a cylinder

Learning objectives

- To calculate the curved surface area of a cylinder
- To calculate the total surface area of a cylinder

A cylinder without a top and bottom is called an **open cylinder**.

When an open cylinder is cut and opened out, it forms a rectangle with the same length as the circumference of the base of the cylinder.

$C = \pi d = 2\pi r$

The **curved surface area** of the cylinder is the same as the area of the rectangle.

The area of the rectangle is $2\pi rh$. So the formula for the curved surface of a cylinder is:

$$A = 2\pi rh$$

The total surface area of the cylinder is the curved surface area plus the area of the circles at each end. The formula for the total surface area of a cylinder is:

$$A = 2\pi rh + 2\pi r^2$$

Example 6

An open cylinder has a length of 8 cm and a diameter of 5 cm.

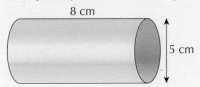

Calculate the area of the curved surface of the cylinder.

Give your answer correct to one decimal place.

The radius of the cylinder is 2.5 cm, so the curved surface area is given by:

$$A = 2\pi rl = 2 \times \pi \times 2.5 \times 8 = 125.7 \text{ cm}^2 \text{ (1 dp)}$$

Example 7

A cylinder has a height of 1.2 m and a radius of 25 cm.

Calculate the total surface area of the cylinder.

Give your answer in square metres, correct to one decimal place.

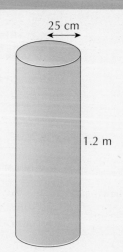

$r = 25$ cm $= 0.25$ m

The total surface area is given by:

$$A = 2\pi rh + 2\pi r^2$$

So $A = 2 \times \pi \times 0.25 \times 1.2 + 2 \times \pi \times 0.25^2 = 2.3 \text{ m}^2 \text{ (1 dp)}$

In this exercise take π = 3.14 or use the 𝛑 key on your calculator.

1. Calculate the total surface area of each cylinder.
   Give your answers correct to one decimal place.

   a

   6·5 m
   3 m

   b

   5 cm
   12 cm

   c

   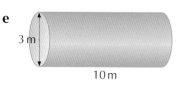
   6 cm
   9 cm

   d

   7 cm
   2 cm

   e

   3 m
   10 m

2. Calculate the curved surface area of each cylinder.
   Give your answers correct to one decimal place.

   a

   3 cm
   15 cm

   b

   1.8 m
   1.2 m

   c

   8.5 cm
   10 cm

3. Calculate the total surface area of a cylinder that has a radius of 4.5 cm and a height of 7.2 cm.

   Give your answer correct to one decimal place.

4. Calculate the curved surface area of a cylinder that has a diameter of 1.2 m and a length of 3.5 m.

   Give your answer correct to one decimal place.

5. A tin of salmon has a diameter of 8.5 cm.

   A label, with a width of 4 cm, goes around the curved surface of the tin.

   Calculate the area of the label, if there is an overlap of 1 cm to glue the ends together.

   Give your answer correct to the nearest square centimetre.

   8.5 cm
   SALMON

6 A cylindrical waste bin has a diameter of
20 cm and a height of 24 cm.

Calculate the total surface area of the bin.

Give your answer correct to the nearest square
centimetre.

 7 A cylindrical tube has a diameter of 8 cm
and a height of 26 cm.

   a Calculate the outside surface area of the tube.

   b The tube has a lid with a diameter of 8.5 cm.

      Calculate the outside surface area of the lid,
      which is 0.8 cm deep.

   c Calculate the total surface area of the tube with
      the lid on.

Give all of your answers correct to one decimal place.

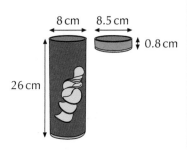

Problem solving: Composite shapes

A A 3D composite shape is made
from a cylinder and a cuboid.

   a Calculate the volume of the shape.

      Give your answer correct to one
      decimal place.

   b Calculate the total surface area
      of the shape.

      Give your answer correct to one
      decimal place.

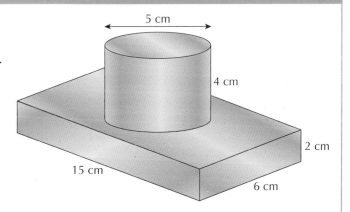

B A plastic cylindrical pipe has a length of 3 m.

The outside diameter is 23 cm and the inside diameter is 20 cm.

   a Calculate the curved surface area of:

     i the outside of the pipe

     ii the inside of the pipe.

      Give your answers in square metres, correct to one decimal place.

   b Calculate the volume of plastic used to make the pipe.

      Give your answer correct to the nearest cubic centimetre.

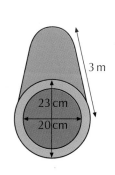

Ready to progress?

I know how to convert the metric units for area, volume and capacity.

I can solve problems, using metric units.

I can calculate the volume of a prism.
I can calculate the surface area of a prism.
I can calculate the volume of a cylinder.
I can calculate the curved surface area of a cylinder.
I can calculate the total surface area of a cylinder.

Review questions

In this exercise take π = 3.14 or use the [π] key on your calculator.

1   Convert each of the units as indicated.

    a   5 m² to cm²   b   7500 mm² to cm²

    c   8 cm³ to mm³   d   5000 cm³ to litres

2   To stay healthy, an adult needs to consume about 1.8 litres of water every day.

    If a glass holds 225 ml, how many glasses of water
    does an adult need to drink every day?

(MR) 3   The diagram shows two circles and a square, ABCD.

    A and B are the centres of the circles. The radius
    of each circle is 6 cm.

    Calculate the area of the shaded part of the square.

    Give your answer correct to one decimal place.

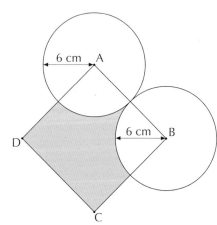

4   a   Work out the volume of this triangular prism.

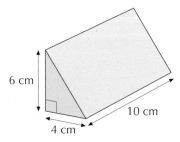

b The cross-section of this prism is made from five identical squares.

Each square has side length 4 cm.

Work out the volume of the prism.

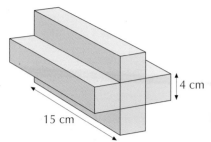

4 cm

15 cm

MR 5 The diagram shows the net for an isosceles triangular prism.

a Use Pythagoras' theorem to work out the length marked x on the diagram.

b Work out the area of one of the isosceles triangles.

c Calculate the total surface area of the prism.

d Calculate the volume of the prism.

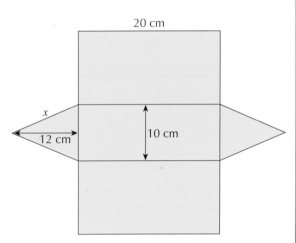

20 cm

x

12 cm

10 cm

6 a Calculate the total surface area of the cylinder.

b Another cylinder has a total surface area that is 15% greater than the first.

Calculate the total surface area of this cylinder.

Give your answers correct to one decimal place.

7 cm

12 cm

PS 7 The four identical cylindrical glasses are completely filled with water. The height of each glass is 12 cm and each has a radius of 3 cm.

The water from the four glasses is then poured into a cylindrical jug, which has a diameter of 14 cm.

a Calculate the volume of water in each glass.

b What is the depth of the water in the jug, after the glasses have been emptied into it?

Give your answers correct to one decimal place.

Problem solving
Packaging cartons of fruit juice

1 Each of these two cartons holds 1 litre of fruit juice.
Both are cuboids, but they have different
dimensions.

Carton A is 9.2 cm by 5.8 cm by 19.4 cm.

Carton B is 7.2 cm by 7.2 cm by 19.8 cm.

  a Calculate the volume of each carton.

  b Which carton has a capacity that is is closer to
  1 litre?

  c Which carton has the smaller total surface area?

  d Which carton is the more efficient to make? Explain why.

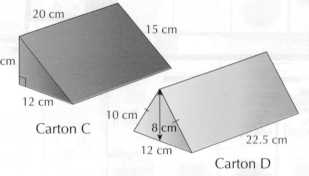

Carton A   Carton B

2 These two triangular prisms could also
hold 1 litre of fruit juice.

  a Calculate the volume of each
  of carton.

  b Which one has the smaller total
  surface area?

  c Would a manufacturer use prisms to
  sell fruit juice? Explain why.

20 cm

15 cm

9 cm

12 cm

Carton C

10 cm

8 cm

12 cm

22.5 cm

Carton D

3 Here are two cylinders that could hold 1 litre
of fruit juice.

  a Calculate the volume of each of carton.

  b Which one has the smaller total surface
  area?

  c Would a manufacturer use cylinders to
  sell fruit juice? Explain why.

8 cm

10 cm

21 cm

14 cm

Taste the
SUNSHINE

FRESH

Carton E   Carton F

4 Here is a cube and a cuboid that could also contain 1 litre of fruit juice.

Work out suitable dimensions for each one, if they both have a volume of 1050 cm³.

Carton G

Carton H

5 Fruit juices are sold in many different types of packaging. Here are some examples.

a Why would manufacturers want to use the container with the smallest surface area?

b Why might the manufacturer choose not to use the carton with the smallest surface area?

c Why are cartons in the shape of cuboids or cylinders so common?

11

Solving equations graphically

This chapter is going to show you:

- how to solve linear equations graphically
- how to draw a quadratic graph
- how to solve quadratic equations graphically
- how to solve simultaneous equations graphically.

You should already know:

- how to draw linear graphs of the form $y = mx + c$
- how to draw a simple quadratic graph.

About this chapter

There are all sorts of different equations that arise from real situations. One example is the satellite dish.

The dish picks up waves and focuses them to a point. It can do this because of its shape. If you were to cut through a satellite dish vertically, across its diameter, you would see that its cross-section is a parabola.

The parabola is the curve formed by the graph of a quadratic equation. This is an equation of the form $y = ax^2 + bx + c$, and the graph always has the same basic shape. You will see parabolic curves used in other sorts of application, such as parabolic headlights.

Some equations are difficult to solve by algebraic methods and are more easily solved by drawing a graph. Such a graph can help you to see roughly what the solution should be and provides a way of checking if your algebraic solution is sensible.

11.1 Graphs from equations in the form $ay \pm bx = c$

Learning objectives

- To draw any linear graph from any linear equation
- To solve a linear equation from a graph

Key word

linear equation

You have already met the **linear equation** $y = mx + c$, which generates a straight-line graph. It is this equation that is seen here as $ay \pm bx = c$. When its graph is plotted, it will, of course, still produce a straight line.

To construct the graph of $ay \pm bx = c$, follow these steps:

- Substitute $x = 0$ into the equation so that it becomes $ay = c$.
- Solve for y, i.e. $y = \dfrac{c}{a}$ so that one point on the graph is $(0, \dfrac{c}{a})$.
- Substitute $y = 0$ into the equation so that it becomes $bx = c$.
- Solve for x, i.e. $x = \dfrac{c}{b}$, so that one point on the graph is $(\dfrac{c}{b}, 0)$.
- Plot the two points on the axes and join them up.

Example 1

Draw the graph of $4y - 5x = 20$.

Substitute $x = 0 \rightarrow 4y = 20$

$\qquad\qquad y = 5$

So the graph passes through $(0, 5)$.

Substitute $y = 0 \rightarrow -5x = 20$

$\qquad\qquad x = -4$

So the graph passes through $(-4, 0)$.

Plot the points and join them up.

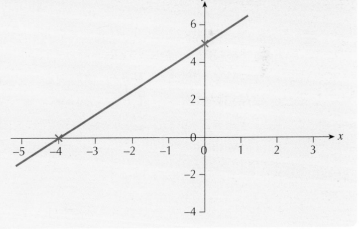

Note that this method is sometimes called the 'cover-up' method as all you have to do to solve for x or y is cover up the other term:

$4y \bigcirc = 20 \rightarrow y = 5$

$\bigcirc - 5x = 20 \rightarrow x = -4$

Example 2

Draw a graph of the equation $y = 5x - 3$.

Use the graph to solve the equation $5x - 3 = 4.5$.

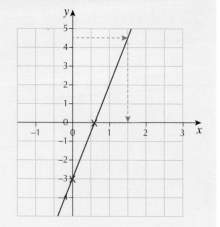

Use the cover-up method to find where the straight line crosses both axes.

Substituting $x = 0 \rightarrow y = -3$

So the graph passes through $(0, -3)$.

Substituting $y = 0 \rightarrow 5x = 3$

$x = \dfrac{3}{5} = 0.6$

So the graph passes through $(0.6, 0)$.

Plot the points and join them up.

Now solve the equation $5x - 3 = 4.5$.

Read from the graph the x-value when $y = 4.5$.

You should see that this is $x = 1.5$.

Exercise 11A

1 Draw the graph of each equation. Use a grid that is numbered from 0 to +10 on both the x-axis and the y-axis.

   a $2y + 3x = 6$ **b** $4y + 3x = 12$ **c** $y + 2x = 8$

   d $3y + 2x = 6$ **e** $5y + 2x = 10$ **f** $2y + 5x = 20$

2 Draw the graph of each equation. Use a grid that is numbered from -10 to $+10$ on both the x-axis and the y-axis.

   a $y - 5x = 10$ **b** $2y - 3x = 12$ **c** $x - 3y = 9$ **d** $2y + 3x = -6$

   e $2y - 5x = 10$ **f** $2x - 3y = 6$ **g** $y + 3x = -9$ **h** $3x - 4y = 12$

   i $y - 2x = 8$ **j** $y + x = -8$ **k** $5y - 2x = 10$ **l** $x - 4y = 10$

3 **a** Draw the graphs of all these equations, on the same grid. Number both axes from -2 to 10.

      i $x + y = 6$ **ii** $x + y = 8$ **iii** $x + y = 10$

      iv $x + y = 2$ **v** $x + y = 1$ **vi** $x + y = 7$

   b What do you notice about all these graphs?

   c Explain how you could now draw the graph of $x + y = -5.3$.

4 **a** Using a grid with axes numbered from -2 to 10, draw the graph of $y = 4x + 3$.

   b Use the graph to solve these equations.

      i $4x + 3 = 7$ **ii** $4x + 3 = 9$ **iii** $4x + 3 = 5$

5 **a** Using a grid with axes numbered from -2 to 10, draw the graph of $y = 5x - 2$.

   b Use the graph to solve these equations.

      i $5x - 2 = 5$ **ii** $5x - 2 = 8.5$ **iii** $5x - 2 = 1.5$

11.2 Graphs from quadratic equations

Learning objective

• To draw graphs from quadratic equations

A **quadratic** equation involves two **variables** where the highest power of one of the variables is a square.

Some examples of quadratic equations are $y = x^2$, $y = x^2 + 3x$, $y = 2x^2 + x - 1$, $y = (x + 2)(x + 1)$.

Example 3

Draw the graph of the equation $y = x^2 + 3x$.

First, draw up a table of values for x, then substitute each value of x into x^2 and $3x$ to determine the y-value.

x	-4	-3	-2	-1	0	1	2
x^2	16	9	4	1	0	1	4
$3x$	-12	-9	-6	-3	0	3	6
$y = x^2 + 3x$	4	0	-2	-2	0	4	10

Now take the pairs of (x, y) coordinates from the table, plot each point on a grid, and join up all the points.

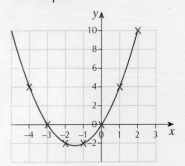

Notice the shape is a smooth curve. It is important always to try to draw a quadratic graph as smoothly as possible, especially at the bottom of the graph where it needs to be a smooth curve.

Example 4

Draw the graph of the equation: $y = x^2 + 2x - 1$.

First, draw up a table of values for x, then substitute for that value of x into x^2 and $3x$ to determine the y value.

x	−4	−3	−2	−1	0	1	2
x^2	16	9	4	1	0	1	4
$2x$	−8	−6	−4	−2	0	2	4
−1	−1	−1	−1	−1	−1	−1	−1
$y = x^2 + 2x - 1$	7	2	−1	−2	−1	2	7

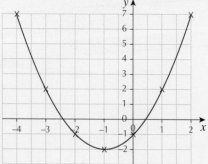

Now take the pairs of (x, y) coordinates from the table, plot each point on a grid, and join up all the points.

Notice that the lowest possible value that y can take is −2, as shown on this graph, but however far the horizontal axis goes, either way, the value of y just keeps on getting larger. Remember that the axes on the graph are just part of infinite number lines.

Exercise 11B

1 **a** Copy and complete this table of values for $y = x^2 + 2x$.

x	−3	−2	−1	0	1	2
x^2	9	4	1	0	1	4
$2x$			−2	0		
$y = x^2 + 2x$				0		

b Draw a grid with the x-axis numbered from −3 to 2 and the y-axis numbered from −2 to 10.

c Use the table to help you draw, on the grid, the graph of $y = x^2 + 2x$.

2 **a** Copy and complete this table of values for $y = x^2 + 4x$.

x	−5	−4	−3	−2	−1	0	1
x^2	25						
$4x$	−20						
$y = x^2 + 4x$	5						

b Draw a grid with the x-axis numbered from −5 to 1 and the y-axis numbered from −5 to 6.

c Use your table to help draw, on the grid, the graph of $y = x^2 + 4x$.

3 **a** Copy and complete this table of values for $y = x^2 + 3x + 2$.

x	−4	−3	−2	−1	0	1
x^2		9			0	
$3x$		−9			0	
2	2	2	2	2	2	2
$y = x^2 + 3x + 2$		2			2	

b Draw a grid with the x-axis numbered from −4 to 2 and the y-axis numbered from −1 to 8.

c Use your table to help you draw, on the grid, the graph of $y = x^2 + 3x - 2$.

4 a Copy and complete this table of values for $y = x^2 + 2x - 3$.

x	−4	−3	−2	−1	0	1	2
x^2							
$2x$							
−3							
$y = x^2 + 2x - 3$							

b Draw a grid with the x-axis numbered from −4 to 2 and the y-axis numbered from −5 to 6.

c Use your table to help you draw, on the grid, the graph of $y = x^2 + 2x - 3$.

5 Construct a table of values for each equation, then, using suitable scales, draw the graphs.

a $y = x^2 + 5x$ **b** $y = x^2 + 3x + 1$ **c** $y = x^2 + 4x - 3$

6 a Copy and complete the table of values for $y = 3x^2 - 5$.

x	−2	−1	0	1	2
x^2	4		0		
$3x^2$	12		0		
$y = 3x^2 - 5$	7		−5		

b Draw a grid with the x-axis numbered from −2 to 2 and the y-axis numbered from −7 to 10.

c Use the table to help you draw, on the grid, the graph of $y = 3x^2 - 5$.

7 a Construct a table of values for each equation, then plot all their graphs on the same pair of axes. Number the x-axis from −4 to 4 and the y-axis from −10 to 55.

     i $y = x^2 - 10$ **ii** $y = 2x^2 - 10$ **iii** $y = 3x^2 - 10$ **iv** $y = 4x^2 - 10$

b Comment on your graphs.

c Sketch onto your diagram the graphs with these equations.

     i $y = \frac{1}{2}x^2 - 10$ **ii** $y = 2\frac{1}{2}x^2 - 10$ **iii** $y = 5x^2 - 10$

8 Andrew makes gold rings.

This chart shows his charges for rings of different diameters.

Diameter (mm)	5	10	15	20	25
Cost (£)	45	61	87	124	170

a Use this information to draw a graph.

b Use your graph to estimate the cost of a ring with diameter:

     i 12 mm   **ii** 24 mm.

c Brian bought a gold ring from Andrew for £100.

     What was the diameter of the ring?

A sledge sliding down a slope travels a distance, d metres, in time, t seconds, where:

$$d = t^2 + 5t$$

A Draw a graph to show the distance covered for the first ten seconds.

B Find the distance travelled after 5.6 seconds.

C Find the time taken to travel 45 m.

11.3 Solving quadratic equations by drawing graphs

Learning objective

• To solve a quadratic equation by drawing a graph

You can find out a lot of information from a quadratic graph, when you know how to do it.

Example 5

a Draw the graph of $y = x^2 + 4x - 5$.

b Use your graph to find the value of y when $x = -3.5$.

c Find the solution to the equation $x^2 + 4x - 5 = 3.2$.

d What are the solutions to the equation $x^2 + 4x - 5 = 0$?

a Draw the graph, as shown.

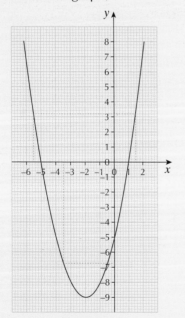

b Draw a dotted line from $x = -3.5$ to the graph.

Following this across to the y-axis.

You can see that when $x = -3.5$, $y = -6.75$.

c To find the solution of the equation $x^2 + 4x - 5 = 3.2$ draw a dotted line across the graph of $y = x^2 + 4x - 5$ where $y = 3.2$.

The dotted line for $y = 3.2$ cuts the graph in two places.

Hence there are two solutions to this equation.

Drawing dotted lines down from the graph to the x-axis, you can see that the solutions are $x = -5.5$ and $x = 1.5$.

d You can find the solution of the equation $x^2 + 4x - 5 = 0$ on the graph where $y = 0$.

This is where the graph cuts the x-axis.

You can see that this will be where $x = -5$ and $x = 1$.

Note:

The solution of a quadratic equation will often give two answers, but not always!

Exercise 11C

For all graphs in this exercise, use a scale of 2 cm to 1 unit on each axis.

1 **a** Draw the graph of $y = x^2$ from $x = -3$ to 3.

    b Write down the value of y when $x = 2.1$.

    c Use the graph to find the solutions to these equations.

       i $x^2 = 3$ **ii** $x^2 = 6$ **iii** $x^2 = 7.5$

2 **a** Draw the graph of $y = x^2 + 2x$ from $x = -3$ to 3.

    b Write down the value of y when $x = 0.7$.

    c Use the graph to find the solutions to these equations.

       i $x^2 + 2x = 2$ **ii** $x^2 + 2x = 1$ **iii** $x^2 + 2x = 0$

3 **a** Draw the graph of $y = x^2 - x$ from $x = -2$ to 3.

    b Write down the value of y when $x = -0.9$.

    c Use the graph to find the solutions to these equations.

       i $x^2 - x = 3$ **ii** $x^2 - x = 1.5$ **iii** $x^2 - x = 0.5$

4 **a** Draw the graph of $y = x^2 + 3x - 2$ from $x = -4$ to 2.

    b Write down the value of y when $x = 1.6$.

    c Use the graph to find the solutions to the following equations:

       i $x^2 + 3x - 2 = 0$ **ii** $x^2 + 3x - 2 = -1$ **iii** $x^2 + 3x = 3$

5 **a** Draw the graph of $y = x^2 + 2x - 3$ from $x = -4$ to 4.

    b Write down the value of y when $x = -1.7$.

    c Use the graph to find the solutions to the following equations:

       i $x^2 + 2x - 3 = 1$ **ii** $x^2 + 2x - 3 = 0$ **iii** $x^2 + 2x - 3 = -1$

6 Draw a graph to find the solutions of $x^2 + x - 7 = 0$.

7 Draw a graph to find the solutions of $x^2 - 4x + 1 = 0$.

Investigation: Can all equations be solved?

A Draw the graph of $y = x^2 + x + 4$.

B You cannot solve the equation $x^2 + x + 4 = 0$.

Explain why it's not possible to solve this equation.

C Write down three more equations that you think you will not be able to find a solution to.

Check each of your equations by drawing the graph.

11.4 Solving simultaneous equations by graphs

Learning objective

• To solve a pair of simultaneous equations

Key word

simultaneous equations

You can draw a graph for any equation. Every pair of coordinates on the graph represents a possible solution for the equation. Hence, when the graphs of two equations are drawn on the same axes, any points where they intercept give a solution that satisfies both equations.

This is known as finding the solution to a pair of **simultaneous equations**.

Example 6

Find the solution of these simultaneous equations by drawing their graphs on the same grid.

$3x + 5y = 25$

$5x - 3y = 2$

First, number the equations.

$3x + 5y = 25$ **(i)**

$5x - 3y = 2$ **(ii)**

Draw the graph of equation **(i)** by plotting the points where the line $3x + 5y = 25$ crosses the axes.

When $x = 0$, $y = 5$, giving the coordinates $(0, 5)$.

When $y = 0$, $x = 8.3$, giving the coordinates $(8.3, 0)$.

Draw the graph of equation **(ii)** by substituting for y, in $5x - 3y = 2$, finding the corresponding value of x and plotting the points. Do this three times.

When $y = 0$, $x = 0.4$, giving the coordinates $(0.4, 0)$.

When $y = 1$, $x = 1$, giving the coordinates $(1, 1)$.

When $y = 6$, $x = 4$, giving the coordinates $(4, 6)$.

The point where the graphs intercept is $(2.5, 3.5)$. So, the solution to the simultaneous equations is $x = 2.5$ and $y = 3.5$.

Exercise 11D

Solve each pair of simultaneous equations by drawing their graphs.

Give your solutions correct to one decimal place.

1 $x + y = 5$
$y = 5x - 4$

2 $x + y = 4$
$2y = 4x - 7$

3 $y = x - 1$
$y = 8 - x$

4 $x + y = 5$
$x = 8 - 3y$

5 $y = 4x - 5$
$x + y = 4$

6 $y = 3x - 5$
$2x + y = 6$

7 $4y = 12 + x$
$3x = 2y - 3$

8 $x + y = 4$
$5y - 3x = 0$

Problem solving: Simultaneous equations

Solve each pair of simultaneous equations by drawing their graphs.

A $y = x^2 - 1$ and $5x + 4y = 20$

B $y = x^2 + x$ and $8x + 3y = 12$

Ready to progress?

I can solve a linear equation from a graph.
I can solve a pair of simultaneous equations by graph.
I can complete a table of values for a quadratic equation and use this to draw a graph of the equation.
I can solve a quadratic equation by drawing a graph.

Review questions

1 a Draw a grid, numbering the axes from −2 to 10.

   Draw the graph of $y = 5x + 4$.

   b Use your graph to solve these equations.
      i $5x + 4 = 11.5$ ii $5x + 4 = -2$ iii $5x + 4 = 15$

2 When a pebble was dropped into a well it fell a distance of d metres in t seconds, where $d = 4.9t^2$.

   a Draw a graph to indicate the distance the pebble fell in the first 5 seconds.
   b The well is 36 m deep. How long will it take the pebble to get to the bottom of the well?

3 a Draw the graph of $y = x^2 - 2x - 4$ for x-values from −3 to 6.

   b Use your graph to find the solutions to these equations.
      i $x^2 - 2x - 4 = 0$ ii $x^2 - 2x - 4 = 1$ iii $x^2 - 2x - 4 = 5$

4 Solve these simultaneous equations by drawing two graphs.

   $x + 4y = 22$

   $y = 4x - 3$

5 Solve the simultaneous equation $y = x^2 + 3$ and $y = 2x + 5$.

6 Todd was finding the volumes of some tins in the kitchen cupboard. He noticed that the heights of all the tins were the same. Their diameters and volumes were as shown in the table. All the values are correct to one decimal place.

Diameter cm	4	5	6	7	8
Volume cm^3	126	196	283	345	503

He wanted to find out the diameter of a tin with the same height that would have a volume of 300 cm^3.

By drawing a suitable graph, find the diameter that would give a volume of 300 cm^3.

(PS) 7 By drawing a suitable graph, find the solution to $(x + 3)(x - 1) = 5$.

(PS) 8 Carl has a ladder that extends to a maximum height of 8 m.

When he sets the ladder up to just below the guttering of the house, the foot of the ladder is 110 cm away from the wall. Approximately how high up the wall is the guttering?

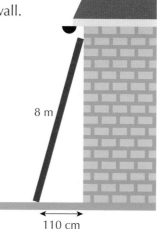

8 m

110 cm

9 a Copy and complete the table of values for $y = x^2 + 3x - 5$.

x	−3	−2	−1	0	1	2
x^2						
$3x$						
−5						
$y = x^2 + 3x - 5$						

b Draw a grid with the x-axis numbered from −3 to 2 and the y-axis numbered from −10 to 10.

c Use the table to help you draw, on the grid, the graph of $y = x^2 + 3x - 5$.

(PS) d Use the graph to solve the equation $x^2 + 3x - 5 = 3$.

(PS) e Use the graph to solve the equation $x^2 + 3x - 5 = 0$.

(MR) f Explain why there is no solution to the equation $x^2 + 3x - 5 = -10$.

(MR) 10 Look what Iwailo and Sunitra are saying.

The equations of the type $y = ax + b$ have solutions for any value of y.

So do equations of the form $y = ax^2 + bx + c$ also have solutions for any value of y?

Answer Sunitra's question, giving reasons.

Challenge
Linear programming

Big industries often call in linear programmers to solve problems for them.
These are mathematicians who use various techniques, such as drawing straight-line graphs from the data they are supplied with, in order to solve problems.
Read how they solve the Widget problem, described here.

A factory that makes Widgets is to be re-equipped.
Two types of machine are used. These are the replacement specifications.

Type	Cost (£)	Daily output	No of people per machine
Cutter	5000	30 widgets	1
Grinder	15 000	40 widgets	2

The company does not wish to spend more than £400 000 on new equipment.
The factory cannot accommodate more than 30 machines.
They do not want to employ more than 50 people to operate the machines.

Problem: How many of each machine should be installed to give the greatest daily output and what is the cost?

The linear programmers go through these stages to solve the problem.

- Call the number of cutters to buy c and the number of grinders g.

- Then the number of machines to purchase is $c + g$.

    So $c + g \leqslant 30$ **1**

- The number of people needed will be $c + 2g$.

    So $c + 2g \leqslant 50$ **2**

- The total cost of the machines will be $5000c + 15\,000g$.

    So $5000c + 15\,000g \leqslant 400\,000$

    Dividing each number by 1000:

    $5c + 15g \leqslant 400$

    Dividing each number by 5:

    $c + 3g \leqslant 80$ **3**

Now, drawing the graph of each of these equations, using $=$ instead of $\leqslant$, will show the region where the solution to the problem can be found.

1 On a suitable grid, draw each graph developed in the calculations above.

   a $c + g = 30$ **b** $c + 2g = 50$ **c** $c + 3g = 80$

2 Look at what each equation represents and shade the regions that do not show any available solutions.

3 You are left with an unshaded region where all possible solutions are available.

   a Find the solution in here that gives the greatest daily output.

   b What is the cost?

12

Compound units

This chapter is going to show you:
- how to solve problems involving speed
- how to calculate and use density
- how to solve problems involving compound units
- how to calculate unit prices and use them to find value for money.

You should already know:
- what is meant by direct proportion
- the standard units of length, mass, capacity and volume
- how to calculate the volumes of simple shapes.

About this chapter

How would you compare how fast athletes can run? How would you compare how quickly water runs out of two taps? How would you describe how a cliff is eroded by the sea?

You measure speed in units such as metres per second or kilometres per hour. You measure the rate of flow of water from a tap or fuel from a pump in litres per second or litres per minute. These are examples of compound units, which are created by combining other units. They often occur when two variables are in direct proportion but, in this chapter, you will look at other examples too. For example, you will consider units that are useful for comparing costs of different-sized packets of goods that you might buy, to compare their value.

12.1 Speed

Learning objective

- To understand and use measures of speed

Jasmine is taking part in a race. She is running at a constant **speed**. She runs 100 metres in 20 seconds.

A formula for working out speed is:

$$\text{speed} = \frac{\text{distance travelled}}{\text{time taken}} \text{ or } \frac{\text{distance}}{\text{time}}$$

Jasmine's speed is $\frac{100}{20} = 5$ metres per second.

The units are metres per second (m/s) because the distance is in metres and the time is in seconds. This is an example of a **compound unit** that involves other units – in this case, metres and seconds.

The units of speed depend on the units used to measure the distance and the time.

Example 1

Work out the speed in each case.

Include units in your answers.

a A car travels 45 km in 30 minutes.

b A runner runs 10 kilometres in 1 hour and 45 minutes.

a $\text{Speed} = \dfrac{\text{distance}}{\text{time}}$

$= \dfrac{45}{30}$

$= 1.5$ km/minute

The units are km/minute because the time is in minutes.

An alternative answer is $\frac{45}{0.5} = 90$ km/h since 30 minutes = 0.5 hours.

b $\text{Speed} = \dfrac{\text{distance}}{\text{time}}$

$\dfrac{10}{1.75} = 5.71$ km/h   One hour 45 minutes is $1\frac{3}{4}$ hours or 1.75 hours.

The answer has been rounded to two decimal places.

An alternative answer is $\frac{10}{105} = 0.095$ km/minute because 1 hour 45 minutes is 105 minutes.

Example 2

A car is travelling at a constant speed of 30 m/s.

a How far does the car travel in one minute?

b How long does the car take to travel one kilometre?

a Speed $= \dfrac{\text{distance}}{\text{time}}$ In this case the speed is 30 m/s and the time is 60 s.

$30 = \dfrac{d}{60}$ The time must be in seconds. Use d for distance.

$30 \times 60 = d$ Multiply by 60 to solve the equation.

$d = 1800$

The distance is 1800 m or 1.8 km. You must include the units.

b Speed $= \dfrac{\text{distance}}{\text{time}}$ In this case the distance is 1000 m.

$30 = \dfrac{1000}{t}$ The distance must be in metres. Use t for time.

$30t = 1000$ First multiply both sides by t.

$t = \dfrac{1000}{30} = 33.333\ldots$ Now divide both sides by 30.

It takes 33 seconds. It is sensible to round the answer to the nearest second.

Exercise 12A

1 A marathon runner runs 40 km in $2\frac{1}{2}$ hours.
Work out his speed, in kilometres per hour (km/h).

2 A train is travelling at a constant speed.

It takes 30 minutes to travel 45 km.

a Work out the speed, in kilometres per minute (km/minute).

b Work out the speed in kilometres per hour (km/h).

3 Calculate the speed in each case. Put units in each answer.

    a Peter runs 320 metres in 50 seconds.   **b** A car travels 15 km in 10 minutes.

    c An aeroplane flies 400 km in half an   **d** A cyclist travels 1500 m in
       hour.   4 minutes.

4 Matthew is cycling at 18 km/h.

Calculate how far he travels in:

    a 1 hour   **b** 4 hours   **c** 1.5 hours   **d** 2 hours and 30 minutes.

5 Paul can swim at a constant speed of 3 km/h.

Calculate how far he can swim in:

    a 1.5 hours   **b** $\frac{1}{2}$ hour   **c** $\frac{1}{4}$ hour   **d** 1 minute.

6 Calculate the distance travelled in each case.

    a Sharon walks at 3 m/s for two minutes.

    b Nathan drives at 80 km/h for 15 minutes.

    c A plane flies at 700 km/h for 4.5 hours.

    d A snail moves at 0.2 m/minute for 150 seconds.

7 The top speed of a sprinter is 8 m/s.

Calculate the time it takes at that speed to sprint:

    a 40 m   **b** 80 m   **c** 50 m   **d** 200 m.

8 Calculate the time taken to travel:

    a 40 km at 120 km/h   **b** 22 km at 4 km/h   **c** 200 m at 5 m/s   **d** 5 km at 4 m/s.

9 Anita is walking in the countryside. She is travelling at 6 km/h.

    a Copy and complete this table.

Time (*t* hours)	0.5	1	1.5	2	2.5
Distance (*d* km)					

    b Work out how long it takes Anita to travel 8 km.

10 An aeroplane is flying at 850 km/h.

    a Calculate how far the aeroplane flies in $3\frac{1}{2}$ hours.

    b Calculate the time the plane takes to fly 5000 km.

11 This graph shows the journey of a car.

    a How far does the car travel in 20 minutes?

    b Explain how you know the car is travelling at a constant speed.

    c Work out the speed of the car, in kilometres/minute.

    d How long does it take the car to travel 50 km?

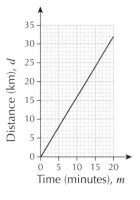

12 A plant grows 3.6 cm in 2 days.

What is the rate of growth in millimetres per hour (mm/h)?

PS **13** The speed of sound is 340 m/s.

There is an explosion 2 km away from Sam.

Calculate how many seconds will pass before Sam hears the explosion.

Challenge: Speed limits

A The national speed limit in the UK is 70 miles per hour (mph).

How fast is that, in metres per second?

Here are some facts that might help you.

A distance of five miles is approximately eight kilometres.

There are 1000 metres in a kilometre.

A mile is 1760 yards.

There are 3600 seconds in an hour.

B In built-up areas the speed limit is 30 miles per hour.

How fast is that, in metres per second?

12.2 More about proportion

Learning objective

• To understand and use density and other compound units

Key words

| density | rate |

In the previous section you learnt about one important example of compound units. Another example is the **rate** of flow, which is a measure of how quickly a liquid is flowing.

Example 3

Water is flowing out of a tap. In 5 minutes, 24 litres flow out of the tap.

a Work out the rate of flow, in litres per minute.

b How long does it take to fill a 7.5 litre bucket?

a The rate of flow is measured as $\dfrac{\text{litres}}{\text{minutes}}$.

This is the number of litres ÷ number of minutes and the unit is litres/minute.

Rate of flow = $\frac{24}{5}$ = 4.8 litres/minute.

b $4.8 = \dfrac{7.5}{m}$ Call the number of minutes m.

  $4.8m = 7.5$ Multiply by m.

  $m = \dfrac{7.5}{4.8} = 1.5625$ Divide by 4.8.

  It takes 1.56 minutes.   Round the answer. It is just over $1\frac{1}{2}$ minutes.

Another example of a compound unit is **density**. This is calculated as 'mass per unit volume':

  density = $\dfrac{\text{mass}}{\text{volume}}$

Examples of possible units are:

• grams per cubic centimetre (g/cm³)

• grams per litre (g/litre).

If you compare equal volumes of different substances, the denser one will be heavier.

Example 4

A piece of iron has a volume of 20 cm³ and a mass of 158 g.

a Calculate the density of iron.   **b** Calculate the mass of 36 cm³ of iron.

a Density = $\dfrac{\text{mass}}{\text{volume}} = \dfrac{158}{20}$

    = 7.9 g/cm³   The units must involve grams and cm³.

b Density = $\dfrac{\text{mass}}{\text{volume}}$

  $7.9 = \dfrac{m}{36}$ Use the answer from part **a**. Use m for the mass.

  $m = 7.9 \times 36 = 284.4$ Multiply by 36.

  The mass is 284 g.

1. A driver takes 20 seconds to put 56 litres of petrol in her fuel tank.
   What is the rate of flow of the petrol? Give units in your answer.

2. A tap is dripping. In 20 minutes 0.3 litres drip from the tap.

   What is the rate of flow, in litres per hour (litres/h)?

3. Water flows down a stream at a rate of 6 litres/s.
   a Calculate how much water flows in one minute.
   b Calculate how much water flows in half an hour.
   c Work out how long it takes for 1000 litres to flow past a particular point.

4. A shower has a rate of flow of 12 litres/minute.
   a Work out how much water is used when Sam has a shower that lasts four minutes.
   b Work out how long it takes to use 1 litre of water.

5. Work out the density of:
   a a piece of metal with a mass of 24 g and a volume of 108 cm^3
   b a piece of wood which has a mass of 400 g and a volume of 320 cm^3
   c a piece of plastic with a mass of 75 g and a volume of 165 cm^3.

6. Balsa wood is used to make model aeroplanes because it has a low density of 0.16 g/cm^3.
   a A piece of balsa wood has a volume of 25 cm^3. Calculate its mass.
   b Calculate the volume of 1 g of balsa wood.

7. Oxygen has a density of 1.43 g/litre.
   Work out the mass of 200 litres of oxygen.

8 Oak has a density of about 0.75 g/cm³.

  a Work out the mass of 1000 cm³ of oak.

  b Work out the volume of a piece of oak with a mass of 150 g.

  c An oak plank measures 3 cm by 15 cm by 120 cm. Work out its mass.

9 Graphite is used to make pencils. It has a density of 2.3 g/cm³.

  a Work out the volume of 50 g of graphite.

  b A pencil contains 1.2 cm³ of graphite. Work out the mass of graphite in 10 pencils.

  c Work out the mass of 100 cm³ of graphite.

10 A piece of copper has a mass of 45 g and a volume of 5 cm³.

  a Work out the density of copper.

  b Work out the volume of 120 g of copper.

  c Work out the mass of a cube of copper with a side of 5 cm.

(PS) 11 The density of gold is 19.3 g/cm³.

  a Work out the mass of 3.5 cm³ of gold.

  b Work out the volume of 3.5 g of gold.

  c The largest gold bar in the world has a mass of 250 kg. Work out its volume.

(PS) 12 This graph shows the connection between mass and volume for a type of steel.

  a Use the graph to find the mass of 2 m³ of steel.

  b Work out the density of the steel. Give your answer in kg/m³.

  c The mass of a particular type of steel beam is 400 kg. Show that the volume of the beam is 0.05 m³.

  d Explain how you know that the mass of the beam is proportional to its volume.

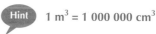
Hint   1 m³ = 1 000 000 cm³

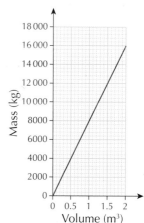

Challenge: Heavy metal

A metal cuboid measures 2 cm by 2 cm by 10 cm. It has a mass of 340 g.

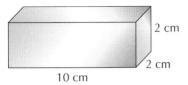

A Calculate the volume of the cuboid.

B Calculate the density of the metal.

C The same metal is used to make a cylinder with a diameter of 3 cm and a length of 6 cm. Calculate the mass of the cylinder.

12.3 Unit costs

Learning objective

• To understand and use unit pricing

Packets and containers of food and other items are sold in different sizes. If you want to compare the prices of the same item, in different sized containers, it is helpful to be able to work out a **unit price**. This is the price of one gram or one litre, or any other suitable unit.

Example 5

Compare the prices of the rice in these packets.

• You could find the cost/gram of rice from each packet.

  For the smaller packet, 89p ÷ 200 = 0.445 p/g (pence per gram).

  For the larger packet, 209p ÷ 500 = 0.418 p/g.

  The larger packet is better value, as it has the lower cost per gram.

• You could find how much you can buy for 1p.

  For the smaller packet, the number of grams/p is 200 ÷ 89 = 2.247 … g/p.

  For the larger packet, it is 500 ÷ 209 = 2.392 … g/p.

  The larger packet is better value because you can buy more for 1p.

This exercise will give you the chance to practise different methods.

> **Note:**
> In shops you will see the weight of an item, given in grams. It is more correct to use the word 'mass' instead of weight. The word 'mass' will be used in this exercise.

Exercise 12C

 1 A bag of pasta has a mass of 250 g and costs 87p.

    a Calculate the cost per 100 g.

    b Calculate the cost per gram.

    c Calculate the number of grams bought for 1p.

    d Calculate the number of grams bought for £1.

 2 A bag of sugar has a mass of 1 kg and costs £2.35.

    a Calculate the cost per 100 g.

    b Calculate the cost per gram.

    c Calculate the number of grams bought for 1p.

    d Calculate the number of grams bought for £1.

 3 Gina buys a 600 g bag of satsumas for £2.40.

    Work out the cost per kilogram.

 4 Raspberries cost £12.50 per kilogram.

    Calculate the mass you can buy for £1.00.

 5 A 160 g can of tuna costs £1.85. A pack of four cans costs £6.20.

    a Work out the cost per 100 g for one can.

    b Work out the cost per 100 g if you buy the pack of four cans.

    c Which is better value? Give a reason for your answer.

6 You can buy tomato puree in tubes or jars. A shop sells 140 g tubes for 69p and 200 g jars for £1.09.

    a Work out the cost per 100 g in a tube of tomato puree.

    b Work out which is better value for money. Justify your answer.

7 This graph shows the price of asparagus.

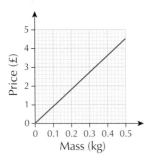

    a Use the graph to find the cost of 400 g of asparagus.

    b Use the graph to find the mass you can buy for £2.00.

    c Explain how the graph shows that cost is proportional to mass.

    d Work out the cost of asparagus, in £/kg.

 8 Aziz buys 700 g of apples for £1.68. Calculate the cost per kilogram.

9 A supermarket sells two cheese selection packs. One has a mass of 490 g and costs £6.00. The other has a mass of 265 g and costs £3.49.

    Calculate the cost per kg of each pack.

10 A shop sells cans of pineapple in two sizes. The prices are shown in this table.

Mass	225 g	435 g
Price	83p	£1.45

    Which is better value? Justify your answer.

11 Milk is sold in litres or pints.

    One pint is 568 ml.

    A supermarket sells 1 litre of milk for 95p or 1 pint for 49p.

    Which is better value? Justify your answer.

(MR) **12** A 600 g box of muesli costs £2.65. An 850 g box costs £3.95.

Which is better value? Justify your answer.

(MR) **13** Toothpaste comes in tubes of different sizes. A 125 ml tube costs £2.99 and a 75 ml tube of the same brand costs £1.89.

Which is better value for money? Justify your answer.

(PS) **14** A 400 g packet of beef costs £4.60.

What price would you expect to pay for a 750 g pack? Justify your answer.

(MR) **15** A pack of four 120 g pots of yogurt costs £2.00.

A 450 g pot of the same yogurt costs £1.79.

Show that the large pot is better value.

(MR) **16** A 350 g block of Cheddar cheese costs £2.90.

A 250 g packet of grated Cheddar cheese costs £1.95.

Show which is better value.

(MR) **17** A 150 g packet of raisins costs £1.09.

A 500 g pack of raisins costs £2.55.

Which is better value? Justify your answer.

Challenge: Value for money

A shop sells cereal in packets of two sizes.

One weighs 350 g and costs £1.79 and the other weighs 575 g and costs £2.85.

They are going to start selling a 750 g pack.

They want it to be slightly better value than the other packs.

They have asked you to recommend a price.

What is your suggestion? Justify your answer.

Ready to progress?

 I can calculate a speed if I know the distance and the time taken.

 I can solve problems involving speed.
I can calculate density and solve problems involving density.
I can solve problems involving other compound units, such as rates of flow.
I can calculate unit prices and use them to find value for money.

Review questions

1 It takes James 8 minutes to walk to the shops. The distance is 400 m.

   a Work out the speed, in metres/minute.

   b How long would it take James to walk 1 km at the same speed?

2 An aeroplane flies at 800 km/h.

   a How far does it fly in $1\frac{1}{2}$ hours?

   b Calculate the time it takes to fly 2000 km.

3 A cheetah can run 100 metres in 4 seconds.
Work out the speed of the cheetah:

   a in m/s

   b in m/min

   c in km/h.

4 This graph shows the distances travelled by a train and a car over a four-hour interval.

   a Work out the speed of the train.

   b Work out the speed of the car.

   c If the train travels d km in t hours, write a formula for d in terms of t.

5 20 litres of water flows out of a tap every minute.
Work out the rate of flow in:

   a litres/second   **b** litres/hour.

6 A paddling pool has a capacity of 200 litres. It is filled at a rate of 25 litres/minute.
Work out how long it takes to fill the pool.

7 This is a block of concrete in the shape of a cuboid.

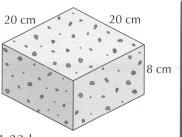

20 cm   20 cm

8 cm

  a Calculate the volume of the block.

  b The density of concrete is 2.4 g/cm^3.

   Calculate the mass of the block of concrete.

8 A piece of metal has a volume of 150 cm^3 and a mass of 1.23 kg.

Calculate the density of the metal in g/cm^3.

9 Densities can be given in g/cm^3 or in kg/m^3.

1 g/cm^3 = 1000 kg/m^3

  a The density of the wood in a tree trunk is 0.85 g/cm^3. Write this in kilograms per cubic metre (kg/m^3).

  b The volume of the tree trunk is 3.8 m^3. Work out the mass of the tree trunk.

 10 A 250 g piece of cheese costs £2.10.

  a Work out the cost per 100 g of the cheese.

  b Work out the cost per kilogram of the cheese.

  c What mass of cheese costs £5.00?

(PS) **11** The cost of petrol is £1.35 per litre. Petrol flows into the tank of a car at a rate of 1.2 litres per second.

Jess puts petrol into her car for half a minute.

  a Work out the quantity of petrol Jess puts in her car.

  b Work out the cost of the petrol.

(PS) **12** A 350 g packet of biscuits costs £1.49. A 450 g packet of the same biscuits costs £1.99.

Which is better value? Justify your answer.

(PS) **13** A box of cereal has a mass of 300 g and costs £2.85.

  a Work out the cost per 100 g.

  b The manufacture decides to give 25% extra for the same price.

   Calculate the new cost per 100 g.

Challenge
Population density

1 Look at the figures in this table.

Country	Area (thousand km^2)	Population (million)
Sweden	410	10
Germany	360	81

The area of Sweden is 410 000 km^2. The population of Sweden is 10 000 000.

  a Write down the area and the population of Germany.

  b The area of Sweden is about 14% larger than the area of Germany.
      What can you say about the populations of the two countries?

For any country, you can work out the population density.

$$\text{Population density} = \frac{\text{population}}{\text{area}}$$

This tells you how crowded a country is. The units are people per km^2.
The population density of Sweden = 10 000 000 ÷ 410 000 = 24 people/km^2.

  c Work out the population density of Germany, to the nearest whole number.

2 This table gives the same information about 11 more European countries.

Country	Area (thousand km^2)	Population (million)
Belgium	31	11
Finland	300	5
France	540	64
Greece	130	11
Ireland	70	5
Italy	300	60
Netherlands	34	17
Poland	310	39
Portugal	92	11
Spain	500	47
United Kingdom	240	64

  a Work out the population density for each country. Give your answers to the
      nearest whole number.

  b If you include Sweden and Germany, you now have data for 13 countries.
      Which country has

     i   the largest population   **ii** the largest population density

     iii the smallest population   **iv** the smallest population density?

3 Draw a set of axes like this on graph paper.

Use a scale of 2 cm to 100 000 km² and 2 cm to 10 million people. You will need values up to 80 million.

Put a cross for each country and label it with the name and population density. Sweden has been done for you.

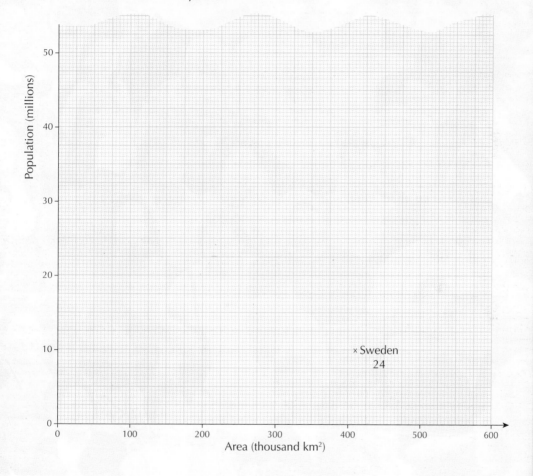

4 Two countries with very different population densities are Singapore and Australia.

Singapore has an area of only 700 km² and a population of 5 400 000.

Australia has an area of 7 700 000 km² and a population of 23 000 000.

a Work out the population density of each of these two countries.

b If you plotted Singapore on your graph, which axis would it be closer to?

c If you extended your graph so that you could plot Australia, which axis would it be closer to?

d Check that this is the case.

13

Right-angled triangles

This chapter is going to show you:

- what trigonometric ratios are and how to recognise them in right-angled triangles
- how to use trigonometry to calculate angles from two known sides in a right-angled triangle
- how to find an unknown length in a right-angled triangle where all angles and one other length are known.

You should already know:

- what a right-angled triangle is
- how to round numbers to a suitable degree of accuracy
- what a hypotenuse is
- how to change a fraction into a decimal.

About this chapter

How could you find the height of a very tall building, or a tree, if all you had was a tape measure, a protractor and a calculator? The answer is, you could use trigonometry!

Trigonometry is the branch of mathematics that studies the relationships between the sides and angles of triangles. In this chapter, you will look at important properties of right-angled triangles and discover relationships that will enable you to calculate the sizes of angles or the lengths of sides of triangles.

You will also be able to work out the height of that tall tree.

13.1 Introducing trigonometric ratios

Learning objective

• To understand what trigonometric ratios are

Key words	
adjacent	cosine
hypotenuse	opposite
similar	sine
tangent	

Similar triangles

Each of these two triangles has angles of 45°, 60° and 75°, although the lengths of the sides are different.

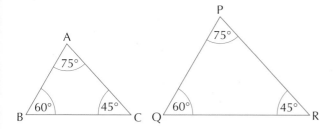

angle A = angle P

angle B = angle Q

angle C = angle R

This means that triangles ABC and PQR are **similar**.

Two triangles are similar if they both have the same set of angles.

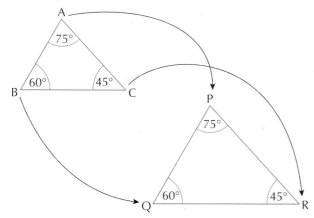

> **Hint** You could say that one triangle is an enlargement of the other.

The lengths of the sides of the triangles do not matter, for the test of similarity. You only need to consider the angles.

Trigonometry in right-angled triangles

You know that every right-angled triangle obeys the rule of Pythagoras' theorem.

In about the fourth or fifth century BC, mathematicians discovered something very interesting about the sides and angles in similar right-angled triangles.

The next exercise will lead you to find out what they discovered.

Work as a class or in small groups for this exercise.

1 Select an angle between 20° and 70°. Each pupil should choose a different angle.

2 **a** Draw six different right-angled triangles, so that the angle you have chosen is at the bottom right, as shown.

> **Hint** If you choose the length of the base to be a whole number of centimetres, the calculations are simpler.

    b Label your triangles A to F.

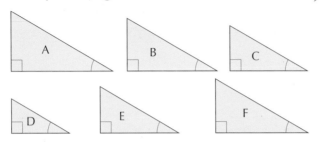

3 **a** Measure, as accurately as you can, the lengths of the sides of all the triangles. Add the lengths as labels on your diagram.

    b Now add extra labels to each triangle, like this.

You already know that the '**hypotenuse**' is the long, sloping side opposite the right angle.

The '**opposite**' is the side across from the angle you are focused on.

The '**adjacent**' is the side next to both the right angle and the angle you are focused on.

> **Hint** You can abbreviate these words to 'hyp', 'opp' and 'adj'.

4 Copy and complete the table below.

Change the fractions in the last three columns to decimals.

Give your answers correct to three decimal places.

Triangle	Opposite	Adjacent	Hypotenuse	Opposite Hypotenuse	Adjacent Hypotenuse	Opposite Adjacent
A						
B						
C						
D						
E						
F						

5 What do you notice?

6 Does this happen with everybody's table with different angles?

7 Now – with a scientific calculator – try the following routines.

    a Key in the angle you have chosen.

    b Press the **sin** key. What do you notice?

    c Key in the angle you have chosen.

    d Press the **cos** key. What do you notice?

    e Key in the angle you have chosen.

    f Press the **tan** key. What do you notice?

    g Does this same result happen for everyone's triangles, whatever the angle chosen?

    h Match up the button **sin**, **cos** and **tan** to

$$\frac{\text{opposite}}{\text{hypotenuse}}, \quad \frac{\text{adjacent}}{\text{hypotenuse}} \quad \text{or} \quad \frac{\text{opposite}}{\text{adjacent}}.$$

Summary

You have discovered that in any right-angled triangle, given the angle x, you can label the sides as 'opposite' (opp), 'adjacent' (adj) and 'hypotenuse' (hyp). But notice that the opposite and adjacent sides are always labelled according to the angle you are looking at.

These thee examples show this.

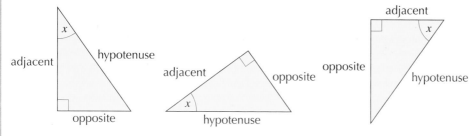

It doesn't matter which way round you see the right-angled triangle. For each angle there is only one adjacent side and one opposite side.

You need to learn and remember these relationships:

sine (sin) $= \dfrac{\text{opposite}}{\text{hypotenuse}}$ **cosine** (cos) $= \dfrac{\text{adjacent}}{\text{hypotenuse}}$ **tangent** (tan) $= \dfrac{\text{opposite}}{\text{adjacent}}$

One way to learn this is to remember a rhyme or a mnemonic, such as:

Silly Old Hitler Couldn't Advance His Troops Over Africa

$S = \dfrac{O}{H} \qquad C = \dfrac{A}{H} \qquad T = \dfrac{O}{A}$

or

Tommy On A   Ship Of His   Caught A Herring

$T = \dfrac{O}{A} \qquad\qquad S = \dfrac{O}{H} \qquad\qquad C = \dfrac{A}{H}$

Make up some of your own mnemonics. Try to use family names or words you will easily remember.

You need to know these ratios by heart.

13.2 How to find trigonometric ratios of angles

Learning objective

- To understand what the trigonometric ratios sine, cosine and tangent are

Now you know what the trigonometric ratios are, you can practise working them out.

Example 1

For each triangle, identify and work out a trigonometric ratio for angle x from the lengths shown.

a

b

c

a Identify 3 as 'opposite', 8 as 'hypotenuse'.

So use SOH.

$\sin x = \dfrac{\text{opp}}{\text{hyp}} = \dfrac{3}{8} = 0.375$

b Identify 5 as 'adjacent', 7 as 'hypotenuse'.

So use CAH.

$\cos x = \dfrac{\text{adj}}{\text{hyp}} = \dfrac{5}{7} = 0.714$

c Identify 8 as 'adjacent', 7 as 'opposite'.

So use TOA.

$\tan x = \dfrac{\text{opp}}{\text{adj}} = \dfrac{7}{8} = 0.875$

Example 2

Sketch a triangle from each trigonometric ratio.

a $\tan x = \dfrac{5}{6}$ **b** $\cos x = \dfrac{2}{7}$ **c** $\sin x = \dfrac{5}{9}$

a $\tan x = \dfrac{5}{6} = \dfrac{\text{opp}}{\text{adj}}$ **b** $\cos x = \dfrac{2}{7} = \dfrac{\text{adj}}{\text{hyp}}$ **c** $\sin x = \dfrac{5}{9} = \dfrac{\text{opp}}{\text{hyp}}$

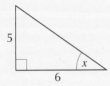

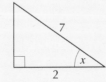

Exercise 13B

1 For each triangle, identify the opposite, the adjacent and the hypotenuse in relation to the angle labelled x.

a

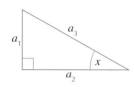

b

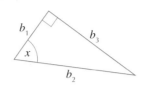

c

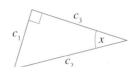

d

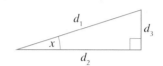

e

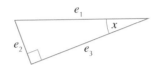

2 For each triangle, copy and complete the trigonometric ratios.
The first one has been done for you.

a

b

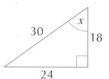

c

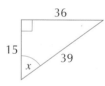

a $\sin x = \dfrac{\text{opp}}{\text{hyp}} = \dfrac{4}{5} = 0.8$

$\cos x = \dfrac{\text{adj}}{\text{hyp}} = \dfrac{3}{5} = \Box$

$\tan x = \dfrac{\text{opp}}{\text{adj}} = \dfrac{4}{3} = \Box$

b $\sin x = \dfrac{\Box}{\Box} = \dfrac{\Box}{\Box} = \Box$

$\cos x = \dfrac{\Box}{\Box} = \dfrac{\Box}{\Box} = \Box$

$\tan x = \dfrac{\Box}{\Box} = \dfrac{\Box}{\Box} = \Box$

c $\sin x = \dfrac{\Box}{\Box} = \dfrac{\Box}{\Box} = \Box$

$\cos x = \dfrac{\Box}{\Box} = \dfrac{\Box}{\Box} = \Box$

$\tan x = \dfrac{\Box}{\Box} = \dfrac{\Box}{\Box} = \Box$

3 For the angle labelled x in each triangle:

i write down which side is 'opp', which is 'adj' and which is 'hyp'

ii write down the fraction for sin, cos and tan.

a

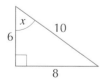

b

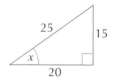

c

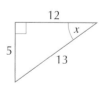

d

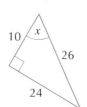

e

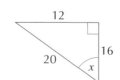

f

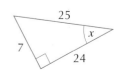

4 For the angle labelled x each triangle:

   i write down which of the trigonometric ratios can be identified

   ii write down the fraction.

  The first one has been done for you.

a 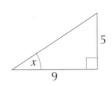   **b** 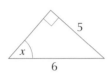   **c**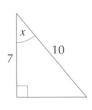

d   **e**   **f**

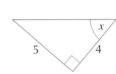

   a $\tan x = \dfrac{\text{opp}}{\text{adj}} = \dfrac{5}{9}$

5 For each part, sketch a right-angled triangle. Label the sides, to illustrate the given facts about the triangle.

  a $\tan x = \dfrac{6}{7}$ **b** $\cos x = \dfrac{3}{5}$ **c** $\sin x = \dfrac{7}{12}$ **d** $\tan x = \dfrac{5}{9}$

  e $\cos x = \dfrac{9}{11}$ **f** $\sin x = \dfrac{7}{11}$ **g** $\cos x = \dfrac{3}{7}$ **h** $\tan x = \dfrac{8}{5}$

6 Look at this triangle.

  a   Write down the fraction for:

     i $\tan x$ **ii** $\sin x$ **iii** $\cos x$.

  b   Write down the fraction for:

     i $\tan y$ **ii** $\sin y$ **iii** $\cos y$.

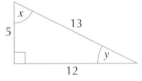

  c   What do you notice about:

     i $\sin x$ and $\cos y$ **ii** $\sin y$ and $\cos x$?

  d   What do you notice about $\tan x$ and $\tan y$?

 7 You are told that $\sin 30° = \dfrac{1}{2}$.

  From this information, sketch three different-sized triangles that each have angles of 90° and 30°.

  Label the angles and the lengths of two appropriate sides.

8 You are told that $\tan 45° = \dfrac{1}{1}$.

  From this information, sketch three different-sized triangles that each have angles of 90° and 45°.

  Label the angles and the lengths of two appropriate sides.

(PS) **(9)** You are told that $\cos 60° = \frac{1}{2}$.

From this information, sketch three different-sized triangles that each have angles of 90° and 60°.

Label the angles and the lengths of two appropriate sides.

Investigation: The sine and tangent of an angle

$\sin 30° = \frac{1}{2}$

A Use Pythagoras' theorem to show that $\tan 30° = \dfrac{1}{\sqrt{3}}$.

B Work out an expression for $\tan 60°$.

C Find an expression for $\sin 60°$.

13.3 Using trigonometric ratios to find angles

Learning objective

- To find the angle identified from a trigonometric ratio

Key word

inverse trigonometric function

From the work you have done so far in this chapter you know that, in a right-angled triangle, if you know the lengths of the hypotenuse and the side opposite an angle, then the angle can only have one value.

But how can you find out what that value is?

Look at this triangle.

The length of the hypotenuse is 8 cm and the length of the side opposite the angle labelled x is 5 cm.

Then:

$$\sin x = \frac{\text{opp}}{\text{hyp}} = \frac{5}{8} = 0.625$$

but how can you identify the angle that has a sine of 0.625?

Fortunately, just as you can use your calculator to change $\frac{5}{8}$ to a decimal, so can you use it to find the angle with a sine of 0.625.

Using your scientific calculator, if you have just divided 5 by 8 to get 0.625, you can go straight on and use the **sin⁻¹** button, which will give the angle of 38.682 187. You would round this to 38.7°.

Note that $\sin^{-1}$ is called an **inverse trigonometric function**, as it allows you to find the angle from the trigonometric function.

Familiarise yourself with all three buttons, **sin⁻¹** **cos⁻¹** and **tan⁻¹**, for the inverse trigonometric functions.

Note

Not all calculators work in this way. Make sure you know how to use these keys on your calculator.

Example 3

A ladder 4 m long leans against a wall. It just reaches a windowsill that is known to be 3.8 m above the ground.

What angle does the ladder make with the ground?

Draw a sketch.

Call the angle at the ground x.

Then you can name the opposite side and the hypotenuse.

Using SOH:

$$\sin = \frac{\text{opp}}{\text{hyp}}$$

$$\sin x = \frac{\text{opp}}{\text{hyp}} = \frac{3.8}{4} = 0.95$$

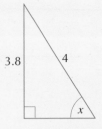

[0] [.] [9] [5] [sin⁻¹] $= 71.8°$

Example 4

A ramp is built for wheelchair access.

It starts at a horizontal distance of 6 m from the step, which is 0.5 m high.

What angle of slope does the ramp have?

Draw a sketch.

Call the angle of the slope x.

Then you can name the opposite and the adjacent sides.

Using TOA:

$$\tan = \frac{\text{opp}}{\text{adj}}$$

$$\tan x = \frac{\text{opp}}{\text{adj}} = \frac{0.5}{6} = 0.083\ 333\ 3$$

[0] [.] [0] [8] [3] [3] [3] [3] [3] [tan⁻¹] $= 4.8°$

Example 5

A builder knew that his ladders were 4.5 metres long when fully extended and that it was in the safest position when the foot of the ladder was 1 metre away from the wall.

What is the 'safe angle' at the floor?

Draw a sketch.

Call the safe angle at the floor x.

Then you know the adjacent side and the hypotenuse.

Using CAH:

$$\cos = \frac{\text{adj}}{\text{hyp}}$$

$$\cos x = \frac{\text{adj}}{\text{hyp}} = \frac{1}{4.5} = 0.222\ 222\ 2$$

[0] [.] [2] [2] [2] [2] [2] [2] [cos⁻¹] $= 77.2°$

Exercise 13C

1 These numbers are all sines of angles. Use your calculator to find out each angle, correct to one decimal place.

   a   0.828   **b**   0.127   **c**   0.632   **d**   0.512

   e   0.395   **f**   0.505   **g**   0.67   **h**   0.99

2 These numbers are all tangents of angles. Use your calculator to find out each angle, correct to one decimal place.

   a   0.641   **b**   0.903   **c**   0.807   **d**   1.56

   e   3.14   **f**   0.845   **g**   5.01   **h**   0.752

3 These numbers are all cosines of angles. Use your calculator to find out each angle, correct to one decimal place.

   a   0.428   **b**   0.705   **c**   0.129   **d**   0.431

   e   0.137   **f**   0.104   **g**   0.811   **h**   0.905

4 These fractions are all sines of angles. Use your calculator to find out each angle, correct to one decimal place.

   a $\dfrac{4}{5}$ **b** $\dfrac{7}{8}$ **c** $\dfrac{1}{9}$ **d** $\dfrac{4}{11}$

   e $\dfrac{1}{7}$ **f** $\dfrac{1}{4}$ **g** $\dfrac{8}{11}$ **h** $\dfrac{9}{13}$

5 These fractions are all tangents of angles. Use your calculator to find out each angle, correct to one decimal place..

   a $\dfrac{6}{7}$ **b** $\dfrac{9}{5}$ **c** $\dfrac{8}{7}$ **d** $\dfrac{1}{6}$

   e $\dfrac{3}{14}$ **f** $\dfrac{18}{17}$ **g** $\dfrac{5}{9}$ **h** $\dfrac{7}{2}$

6 These fractions are all cosines of angles. Use your calculator to find out each angle, correct to one decimal place.

   a $\dfrac{4}{7}$ **b** $\dfrac{7}{15}$ **c** $\dfrac{1}{9}$ **d** $\dfrac{4}{13}$

   e $\dfrac{13}{17}$ **f** $\dfrac{1}{8}$ **g** $\dfrac{8}{11}$ **h** $\dfrac{9}{25}$

7 Calculate the size of the angle labelled x in each triangle. Give your answers correct to one decimal place.

a

b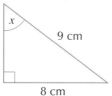

c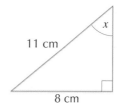

d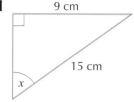

8 Calculate the size of the angle labelled x in each triangle. Give your answers correct to one decimal place.

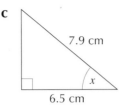

a 11 cm, 13 cm, x

b 9 cm, 7.5 cm, x

c 7.9 cm, 6.5 cm, x

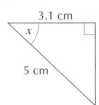

d 3.1 cm, 5 cm, x

9 Calculate the size of the angle labelled x in each triangle. Give your answers correct to one decimal place.

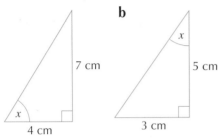

a 7 cm, 4 cm, x

b 5 cm, 3 cm, x

c 11 cm, 17 cm, x

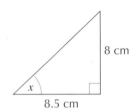

d 12 cm, 14 cm, x

10 Calculate the size of the angle labelled x in each triangle. Give your answers correct to one decimal place.

a 6 cm, 7 cm, x

b 6 cm, 10 cm, x

c 12 cm, 14 cm, x

d 8 cm, 8.5 cm, x

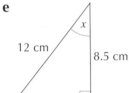

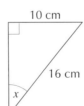

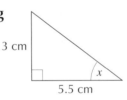

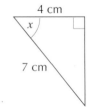

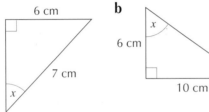

e 12 cm, 8.5 cm, x

f 10 cm, 16 cm, x

g 3 cm, 5.5 cm, x

h 4 cm, 7 cm, x

PS **11** When sliding down a 400 m slope, a skier dropped by a vertical height of 320 m.

At what angle is the slope to the horizontal?

PS **12** A ladder 4.8 m long rests against a wall. It reaches 4 m up the wall. At what angle is the ladder leaning against the wall?

PS **13** Ewan is 1.5m tall. He stood on level ground, looking up at the top of a tower that he knew to be 80 m tall. He was 200 metres away from the bottom of the tower.

What angle from the horizontal did he need to look up, to see the top of the tower?

Investigation: Squares of sine and cosine

A Choose any angle, $x°$.

B Calculate the value of $(\cos x°)^2 + (\sin x°)^2$.

C Repeat steps **A** and **B** for some different angles.

D What do you notice?

E Do you think this will always happen?

13.4 Using trigonometric ratios to find lengths

Learning objective

- To find an unknown length of a right-angled triangle given one side and another angle

Now that you know how to find the sin, cos or tan of any angle, if you are told an angle in a right-angled triangle, as well as an appropriate length, then you can calculate the lengths of the other sides.

Example 6

Find the length of the side labelled x in the diagram.

You should see that the opposite (x), the hypotenuse (6) and the angle 62° are all labelled.

Remember that:

$$\sin 62° = \frac{\text{opp}}{\text{hyp}} = \frac{x}{6}$$

Rearranging, you can write:

$$\frac{x}{6} = \sin 62°$$

Then $x = 6 \times \sin 62°$ Multiply both sides by 6.

This is calculated on most calculators as:

So $x = 5.3$ cm (1 dp).

> **Hint** Check exactly how to use trigonometric functions on your calculator.

Example 7

Find the length of the side labelled x in the diagram.

The adjacent (x), the hypotenuse (15) and the angle 27° are all labelled.

Remember that:

$$\cos 27° = \frac{\text{adj}}{\text{hyp}} = \frac{x}{15}$$

(*Continued*)

So:

$$\frac{x}{15} = \cos 27°$$

Then $x = 15 \times \cos 27°$ Multiply both sides by 15.

This is calculated on most calculators as:

1 5 × 2 7 cos =

So $x = 13.4$ cm (1 dp).

Example 8

Find the length of the side labelled x in the diagram.

The opposite (x), the adjacent (7) and the angle 70° are all labelled.

Remember that:

$$\tan 70° = \frac{\text{opp}}{\text{adj}} = \frac{x}{7}$$

So:

$$\frac{x}{7} = \tan 70°$$

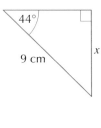

Then $x = 7 \times \tan 70°$ Multiply both sides by 7.

This is calculated on most calculators as:

7 × 7 0 tan =

So $x = 19.2$ cm (1 dp).

Exercise 13D

1. Calculate the length labelled x in each right-angled triangle. Give your answers correct to one decimal place.

a 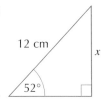   **b** 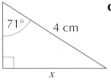   **c** 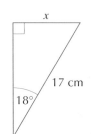   **d**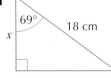

2. Calculate the length labelled x in each right-angled triangle. Give your answers correct to one decimal place.

a 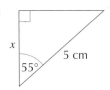   **b**   **c** 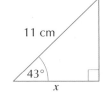   **d**

3 Calculate the length labelled x in each right-angled triangle. Give your answers correct to one decimal place.

a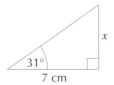

b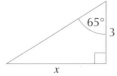

c

d

e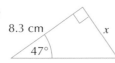

4 Calculate the length labelled x in each right-angled triangle. Give your answers correct to one decimal place.

a

b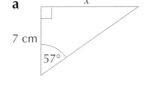

c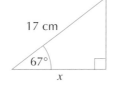

d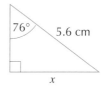

e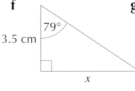

f

g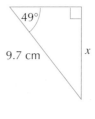

h

(PS) **5** A ship sails on a direction of N75°E for 150 km.

    a Draw a sketch of the ships journey, showing a north line and the east direction.

    b Use your diagram to help you find how far east the ship has sailed.

(PS) **6** A plank 8 metres long is leaning against a wall at an angle of 30° with the horizontal. How far up the wall does the plank reach?

(PS) **7** A right-angled triangle has another angle of 50° and the length of its hypotenuse is 5 cm. Calculate the area of the triangle.

(MR) **8** Explain why sin 45° = cos 45°.

(PS) **9** You fly for 300 km on a bearing of 150°.

    How far: **a** east have you flown   **b** south have you flown?

Investigation: Ratio relations

A Choose any angle and find its sine, cosine and tangent.

B There is a simple way of combining any two of these values to give the other one! Find the rule that connects tan, sin and cos.

Hint Try adding, subtracting, multiplying and dividing.

C Will this always work?

Ready to progress?

I understand what trigonometry is and can identify the sine, cosine and tangent of angles in right-angled triangles.
Given two sides of a right-angled triangle, I can calculate the other angles in the triangle.
Given the angles of a triangle and the length of one side, I can calculate the lengths of the other sides.

Review questions

1   For the marked angle x in each triangle, write down the fractions representing the sine, cosine and tangent.

a

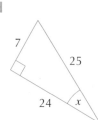

b

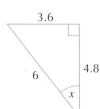

c

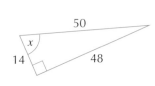

d

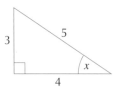

e

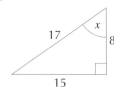

f

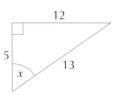

2   For each triangle, calculate the size of the angle labelled x. Give your answers correct to one decimal place.

a

b

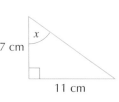

c

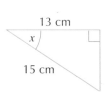

d

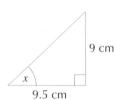

d

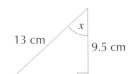

e

f

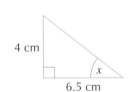

g

3   Calculate the area of each shape.

a

67°

8 cm

b

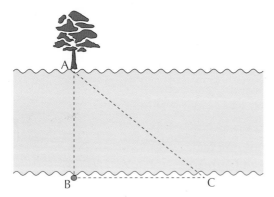

50°

15 cm

7 cm

 4   An isosceles triangle has two sides each 9 cm long and one side of 5 cm.

   a   What is the size of the two identical angles?

   b   What is the area of the triangle?

5   If $\tan x = \frac{4}{3}$ and $\sin x = \frac{4}{5}$, write down the value of $\cos x$.

 6   Lewis is asked to find out how wide a river is.

He places a stone at a point B, directly opposite a tree at A on the river bank on the other side of the river. He then walks 50 m along the bank of the river to a point C, and estimates that the angle ACB is 20°.

Complete his calculation of the width of the river.

7   Khalid is 160 cm tall.

He casts a shadow of length 50 cm.

   a   Calculate the angle that the rays of the Sun make with the horizontal at this time.

   b   At the same time, the shadow of a building is 23 metres long.
       What is the height of the building?

Investigation

Barnes Wallis and the bouncing bomb

In 1943, during the World War 2, Barnes Wallis developed a bomb that would bounce along the water in a dam and explode against the wall holding the water back. He developed this bomb because he thought it was the best way to stop the industry in Germany at that time, while hurting the least number of people.

A Lancaster Bomber was modified to enable it to deliver these bombs to The Mohne Dam in the Ruhr Valley of Germany.

The converging lights

For the bouncing bomb to work it had to be released from the aeroplane at exactly the right height (60 feet). To achieve this accuracy, two angled spotlights were mounted on the front and the rear of the aeroplane. When the two spotlights converged, the pilot knew that the aeroplane was at exactly the right height.

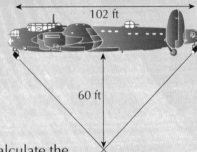

1   Sketch the right-angled triangle that you can use to calculate the angle at which each light must be set from the underside of the aeroplane.

2   Use trigonometry to calculate this angle.

3   Suppose a different aeroplane were used, that was 85 feet long. What angle would the lights need to be angled at?

4   The pilot of an aeroplane that was 70 feet long wished to know when he was exactly 100 feet above a target. What angle would he need to set similar lights to achieve this?

The bomb sights

Not only had the bomb to be dropped from the aeroplane when it was at the right height, it also had to be at the right distance from the dam wall (389 metres).

To get this part of the operation right, two nails were mounted on a sight guide. Then the bomb had to be released at the precise moment when the two nails were seen to be exactly in line with the twin turrets of the dam.

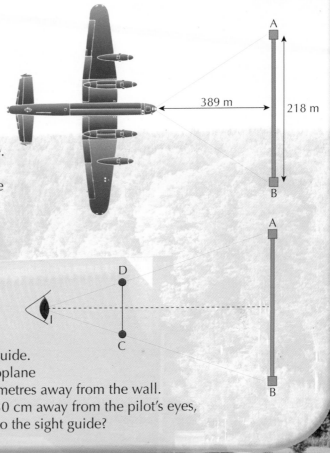

389 m

218 m

5 In the top diagram, A and B are the positions of the dam turrets. What is the angle made at the aeroplane where the lines from each turret meet?

6 In the bottom diagram, D and C are the position of the nails on the sight guide. They help the pilot to line up the aeroplane and to know when he is exactly 389 metres away from the wall. If the nails on the sight are on a line 30 cm away from the pilot's eyes, how far apart must the nails be set into the sight guide?

Durchbruch der Möhnetalsperre

14

Revision and GCSE preparation

This chapter is going to:

- help you to practise and revise topics covered in your current course
- get you started on your GCSE course.

Practice

Practice in rules of algebra and solving equations

Exercise 14A

1 **a** What is the next pair of coordinates in this list?

(2, 1), (4, 3), (6, 5), (8, 7), ...

b Explain why the coordinates (29, 28) could not be part of this sequence.

2 Expand the brackets and simplify each expression if possible.

a $4(x - 5)$ **b** $3(2x + 1) + 5x$ **c** $3(x - 2) + 2(x + 4)$

d $5(3x + 4) + 2(x - 2)$ **e** $4(2x + 1) - 3(x - 6)$

3 Two friends, Selma and Khalid, are revising algebra.

Selma says: 'I am thinking of a number. If you multiply it by 6 and add 3 you get an answer of 12.'

Khalid says: 'I am thinking of a number. If you multiply it by 3 and subtract 6 you get the same answer as adding the number to 7.'

a Call Selma's number x and form an equation. Then solve the equation.

b Call Khalid's number y and form an equation. Then solve the equation.

4 Solve each equation.

a $3x + 7 = x + 10$ **b** $5x - 6 = 10 - 3x$ **c** $3(x + 3) = x + 8$

5 Look at the algebraic expressions on these cards.

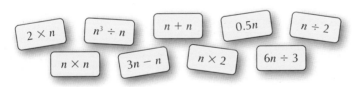

a Which two expressions will always give the same answer as $\frac{n}{2}$?

b Which two expressions will always give the same answer as n^2?

c Five of the expressions are the same as $2n$.

Write an expression of your own that is the same as $2n$.

6 **a** Two of the expressions below are equivalent. Which ones are they?

$3(4x - 6)$ $2(6x - 4)$ $12(x - 3)$ $6(2x - 3)$ $8(4x - 1)$

b Factorise this expression.

$6y - 12$

c Factorise this expression as fully as possible.

$9y^2 - 6y$

Practice in graphs

Exercise 14B

You will need graph paper or centimetre-squared paper for this exercise.

Axes numbered $-6 \leq x \leq 6$ and $-6 \leq y \leq 6$ will be big enough for any graphs you are asked to draw.

1 Draw and label the graph of each equation.

    a $y = 2x + 1$ **b** $y = \frac{1}{2}x - 1$ **c** $x + y = 3$

(MR) 2 Does the point (20, 30) lie on the line $y = 2x - 10$?

    Explain your answer.

(PS) 3 The distance–time graph shows the journey of a cross-country runner on a 5-kilometre practice run.

At one point she stopped to get her breath back and at another point she ran through wet grass.

    a   For how long did she stop to get her breath back?

    b   Explain how you know which part of the graph represents her run through wet grass.

 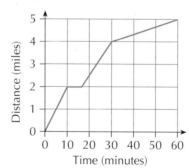

(PS) 4 In a house, the hot-water tank automatically refills with cold water whenever hot water is taken out. The heating system then reheats the water to the pre-set temperature.

Mum always has a shower in the morning.

Eve, the youngest daughter, always has a bath.

The two teenage children get up so late that they only have time to wash their hands and faces.

The graph shows the temperature of the water in the hot water tank between 7:00 am and 9:00 am one morning.

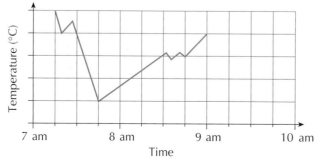

a At what time did Mum have her shower?

b At what time did Eve have her bath?

c At what time did the teenagers wash?

d Grandad likes to have as hot a bath as possible, once everyone else has left the house at 9:00 am. Estimate at what time the water will be back to its maximum temperature.

5 The graph shows the line $y = 2x + 2$.

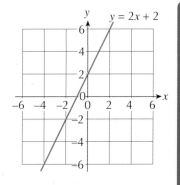

a Copy the graph and draw and label the line $y = 2x - 1$ on the same axes.

b Draw and label the line $y = x + 2$ on the same axes.

c Write down the coordinates of the point where the graphs $y = 2x - 1$ and $y = x + 2$ intersect.

6 The diagram shows a rectangle ABCD.

a The equation of the line AB is $x = 1$. What is the equation of the line CB?

b The equation of the diagonal AC is $y = 2x + 2$. What is the equation of the diagonal BD?

c Write down the equations of the two lines of symmetry of the rectangle.

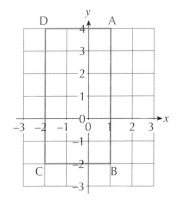

7 Here are the equations of six graphs.

A $y = 2x - 1$ **B** $y = x$ **C** $y = 2$ **D** $x = 2$ **E** $y = 2x + 3$ **F** $y = \frac{1}{2}x - 1$

a Which two graphs are parallel?

b Which two graphs are perpendicular?

c Which pair of graphs cross the y-axis at the same point?

d Which pair of graphs cross the x-axis at the same point?

Practice in geometry and measures

Exercise 14C

1　**a**　Describe angles A–E in the diagram. Choose the correct words from the box.

acute	obtuse	reflex	right-angled

　　b　Is angle A bigger, smaller or the same size as angle C? Explain your answer.

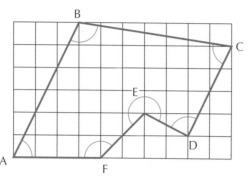

2　**a**　Copy and complete the two-way table to show the symmetries of each shape.
　　　　Shape A has been done for you.

		Number of lines of symmetry				
		0	1	2	3	4
Order of rotational symmetry	1	A				
	2					
	3					
	4					

　　b　Name a quadrilateral that has two lines of symmetry and rotational symmetry of order 2.

3 For each shape, find: **i** the perimeter   **ii** the area.

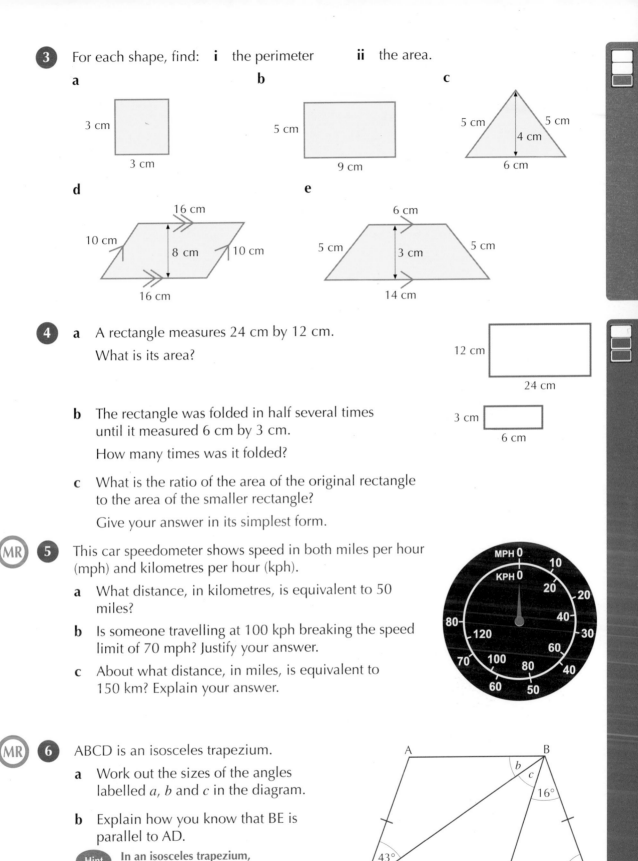

a

3 cm

3 cm

b

5 cm

9 cm

c

5 cm   5 cm

4 cm

6 cm

d

16 cm

10 cm

8 cm   10 cm

16 cm

e

6 cm

5 cm   3 cm   5 cm

14 cm

4   **a**   A rectangle measures 24 cm by 12 cm.

What is its area?

12 cm

24 cm

b   The rectangle was folded in half several times until it measured 6 cm by 3 cm.

How many times was it folded?

3 cm

6 cm

c   What is the ratio of the area of the original rectangle to the area of the smaller rectangle?

Give your answer in its simplest form.

MR **5**   This car speedometer shows speed in both miles per hour (mph) and kilometres per hour (kph).

a   What distance, in kilometres, is equivalent to 50 miles?

b   Is someone travelling at 100 kph breaking the speed limit of 70 mph? Justify your answer.

c   About what distance, in miles, is equivalent to 150 km? Explain your answer.

MR **6**   ABCD is an isosceles trapezium.

a   Work out the sizes of the angles labelled a, b and c in the diagram.

b   Explain how you know that BE is parallel to AD.

Hint   In an isosceles trapezium, the sloping sides are equal and the base angles are equal.

7 For each circle, calculate:

i the circumference   **ii** the area.

Take π = 3.14 or use the π key on your calculator. Give your answers correct to one decimal place.

a
3 cm

b
4.5 cm

c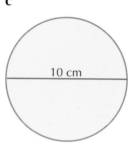
10 cm

d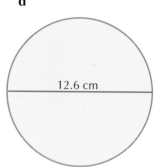
12.6 cm

8 For each 3D shape, calculate:

i the total surface area   **ii** the volume.

a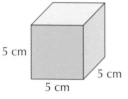
5 cm
5 cm
5 cm

b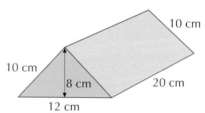
2 cm
10 cm
5 cm

c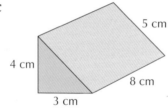
5 cm
4 cm
8 cm
3 cm

d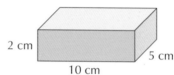
10 cm
10 cm
8 cm
20 cm
12 cm

9 The diagram shows a cuboid and a triangular prism.

Both solids have the same volume.

Use this information to calculate the length of the prism.

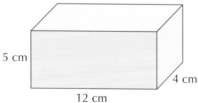
5 cm
4 cm
12 cm
4 cm
3 cm
l

10 **a** Calculate the length of the side marked *x* in this right-angled triangle. Give your answer correct to one decimal place.

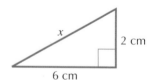
x
2 cm
6 cm

b Calculate the length of the side marked *y* in this right-angled triangle. Give your answer correct to one decimal place.

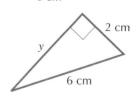
2 cm
y
6 cm

11 Calculate the area of the square drawn on the centimetre grid.

PS **12** A circle has a circumference of 20 cm.

    a Calculate the diameter of the circle.

    b Calculate the area of the circle.

Take $\pi = 3.14$ or use the π key on your calculator. Give each answer correct to one decimal place.

13 Calculate the perimeter and the area of the shape below.

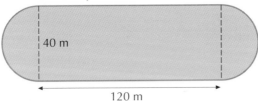

40 m

120 m

Take $\pi = 3.14$ or use the π key on your calculator. Give your answers correct to three significant figures.

PS **14** A coin is stamped from a square sheet of metal.

Investigate the percentage waste for coins of different sizes.

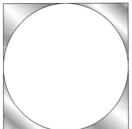

Practice in statistics

Exercise 14D

1 Hakim has five cards.

a What is the mode of the numbers on the cards?

b What is the median of the numbers on the cards?

c What is the mean of the numbers on the cards?

2 Tess throws two four-sided dice, each numbered 1, 2, 3, 4. The table shows all the possible total scores.

		Score on first dice			
		1	2	3	4
Score on second dice	1	2	3	4	5
	2	3	4	5	6
	3	4	5	6	7
	4	5	6	7	8

a When the two dice are thrown, what is the probability that the total score is a square number?

b When the two dice are thrown, what is the probability that the score is greater than 5?

c **i** Draw a table to show all the possible products, if the numbers on each of the dice are multiplied together.

   ii What is the probability that the product is a number less than 17?

3 These are Paul's marks for his last nine mathematics homework tasks.

9, 3, 5, 4, 4, 7, 5, 8, 6

a What is the range of his marks?

b What is the median mark?

c After checking the final homework, Paul realised that his teacher did not mark one of the questions. Once this had been marked, Paul's mark increased from 6 to 8.

Say whether each of the statements, **i**, **ii** and **iii** is true, false or if it is not possible to say.

Explain your answers.

i The mode of the marks has increased.

ii The median mark has increased.

iii The mean mark has increased.

4 The probability that a ball taken at random from a bag is black is 0.7. What is the probability that a ball taken at random from the same bag is not black?

(MR) **5** A school quiz team is made up of pupils from four different classes. The table shows the numbers of pupils in the team from each class.

Class	Number of pupils
A	4
B	3
C	8
D	5

a Represent this information in a pie chart.

b Holly says: 'The percentage of pupils chosen from class C is double the percentage chosen from class A.'

Explain why this might not be true.

6 Ishmael is always late, on time or early for school. The probability that he is late is 0.1 and the probability that he is on time is 0.3.

a What is the probability that he is late or on time?

b What is the probability that he is early?

7 A group of 50 pupils are told to draw two straight lines on a piece of paper. Seven pupils draw parallel lines, twelve draw perpendicular lines and the rest draw lines that are neither parallel nor perpendicular.

Use these results to estimate the probability that a pupil chosen at random has:

a drawn parallel lines

b drawn perpendicular lines

c drawn lines that are neither parallel nor perpendicular.

(MR) **8** A five-sided spinner was spun 50 times. These are the results.

Number on spinner	1	2	3	4	5
Frequency	8	11	10	6	15

a Write down the experimental probability of the spinner landing on the number 4.

b Write down the theoretical probability of a fair, five-sided spinner landing on the number 4.

c Compare the experimental and theoretical probabilities and say whether you think the spinner is fair.

d How many fours would you expect if the spinner were spun 250 times?

9 Lee and Alex are planning a survey of what pupils at their school prefer to do at the local entertainment complex, where there is a cinema, a bowling alley, a games arcade and a disco.

    a Alex decides to give out a questionnaire to all the pupils in a Year 7 tutor group.

       Explain why this may not give reliable results for the survey.

    b Lee decides to include this question in his questionnaire:

> How many times in a week do you go to the entertainment complex?
>
> Never ☐   1–2 times ☐   2–5 times ☐   every day ☐

       Explain why this is not a good question.

  10 x is a whole number bigger than 1.

3x   2x + 1   x + 1   3x + 2   6x + 1

For the values on the five cards:

    a work out the median value   **b** calculate the mean value.

11 The scatter diagram shows the value and the mileage for a number of cars. (The mileage is the total distance the car has travelled since new.)

The value of each car is given as a percentage of its value when it was new.

A line of best fit has been drawn on the scatter diagram.

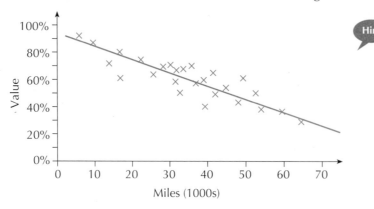

> **Hint** A line of best fit is drawn so that there are as many values above it as below it. It shows the trend of the data.

    a What does the scatter diagram show about the relationship between the value of a car and its mileage?

    b A car has a mileage of 45 000. Estimate its value, as a percentage of its value when new.

    c A car cost £12 000 when it was new. It is now worth £7800. Use this information to estimate how many miles it has travelled.

12 These are the times taken (T seconds) by 20 pupils to run 100 m.

Boys	13.1	14.0	17.9	15.2	15.9	17.5	13.9	21.3	15.5	17.6
Girls	15.3	17.8	16.3	18.0	19.2	21.4	13.5	18.2	18.4	13.6

a Copy and complete the two-way table to show the frequencies.

		Boys	Girls
Frequency	$12 \leqslant T < 14$		
	$14 \leqslant T < 16$		
	$16 \leqslant T < 18$		
	$18 \leqslant T < 20$		
	$20 \leqslant T < 22$		

b What percentage completed the 100 m in less than 16 seconds?

c Which is the modal class for the girls?

d In which class is the median time for the boys?

(MR) 13 It was estimated that there were 58 836 700 people living in Klingdom in mid-2014.

This was an increase of 1.4 million people (2.4 per cent) since 2004.

The graph shows the population (in thousands) of Klingdom from 2004 to 2014.

Explain why it may be misleading.

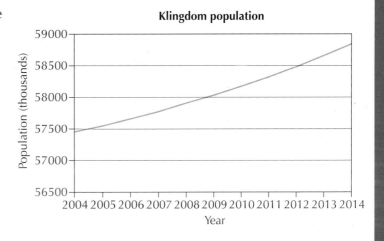

Klingdom population

14 State whether the outcomes in each pair are mutually exclusive or not mutually exclusive. Explain your answers.

a An ordinary, six-sided dice landing on an even number.

The dice landing on a prime number.

b Two coins being thrown and getting at least one head.

The two coins being thrown and getting two tails.

c Two coins being thrown and getting at least one tail.

The two coins being thrown and getting two tails.

Practice in number problems

Exercise 14E

Work out each of these. Show your working.

Check your answers with a calculator afterwards.

1 A typist can type 54 words per minute on average.

How many words can he type in 15 minutes?

(FS) 2 Small chocolate eggs cost 43p each.

Mrs Owen wants to buy an egg for each of her class of 28 pupils.

How much will this cost her?

3 There are 972 pupils in a school. Each tutor group has 27 pupils in it.

How many tutor groups are there?

4 In a road-race, there were 2200 entrants.

a To get them to the start, the organisers used a fleet of 52-seater buses.

How many buses were needed?

b The race was 15 miles long and all the entrants completed the course.

How many miles, in total, did all the runners cover?

(FS) 5 At a school fair, cups of tea were 32p each. The school sold 182 cups.

a How much money did they take?

b The school used plastic cups that came in packs of 25.

They bought 24 packs.

How many cups were left over?

6 a A cinema has 37 rows of seats. Each row contains 22 seats.

How many people can sit in the cinema altogether?

(FS) b Tuesday is 'all seats one price' night. There were 220 customers who paid a total of £572.

What was the cost of one seat?

7 A library has 700 books to distribute equally among 12 local schools.

a How many books will each school get?

b The library keeps any books left over. How many books is this?

8 The label on the side of a 1.5 kg cereal box says that there are 66 g of carbohydrate in a 100 g portion.

How many grams of carbohydrate will Dan consume if he eats the whole box at once?

(FS) 9 A first-class stamp costs 60p and a second-class stamp costs 50p.

How much does it cost to send 63 letters first class and 78 letters second class?

12 members of a running club hire a minivan to do the Three Peaks race (climbing the highest mountains in England, Scotland and Wales). The van costs £25 per day plus 12p per mile. The van uses a litre of petrol for every 6 miles travelled. Petrol costs £1.30 per litre. The van is hired for three days and the total distance covered is 1500 miles.

a How much does it cost to hire the van?

b How many litres of petrol are used?

c If the total cost is shared equally how much does each member pay?

Revision

Revision of equivalent fractions

Example 1

Cancel each fraction to its lowest terms.

a $\dfrac{9}{30}$ **b** $\dfrac{8}{18}$ **c** $\dfrac{15}{55}$

To cancel a fraction you need to find the highest common factor (HCF) of the numerator and the denominator, then divide both by it.

a The HCF of 9 and 30 is 3. $\dfrac{9}{30} = \dfrac{9 \div 3}{30 \div 3} = \dfrac{3}{10}$

b The HCF of 8 and 18 is 2. $\dfrac{8}{18} = \dfrac{\overset{4}{\cancel{8}}}{\underset{9}{\cancel{18}}} = \dfrac{4}{9}$

c The HCF of 15 and 55 is 5.

$\dfrac{15}{55} = \dfrac{\overset{3}{\cancel{15}}}{\underset{11}{\cancel{55}}} = \dfrac{3}{11}$

Example 2

Which of the fractions below is closest to $\frac{1}{2}$?

$\dfrac{3}{5}$ $\dfrac{4}{9}$ $\dfrac{6}{11}$ $\dfrac{5}{12}$

It would be very difficult to find a common denominator, so convert all the fractions to decimals.

$\dfrac{3}{5} = 0.6$ $\dfrac{4}{9} = 0.444$ $\dfrac{6}{11} = 0.545$ $\dfrac{5}{12} = 0.417$

Of these $\frac{6}{11}$ is closest to $\frac{1}{2}$ (0.5).

1 Cancel each fraction to its lowest terms.

a $\frac{9}{12}$ **b** $\frac{15}{25}$ **c** $\frac{7}{21}$ **d** $\frac{9}{15}$

e $\frac{14}{35}$ **f** $\frac{16}{40}$ **g** $\frac{12}{30}$ **h** $\frac{18}{24}$

2 Copy these equivalent fractions and fill in the missing numbers.

a $\frac{5}{15} = \frac{\square}{45}$ **b** $\frac{6}{21} = \frac{\square}{7}$ **c** $\frac{12}{21} = \frac{4}{\square}$ **d** $\frac{8}{28} = \frac{\square}{35}$

e $\frac{12}{15} = \frac{\square}{25}$ **f** $\frac{9}{24} = \frac{\square}{16}$ **g** $\frac{18}{30} = \frac{12}{\square}$ **h** $\frac{10}{35} = \frac{\square}{21}$

(FS) 3 Calculate each amount.

a $\frac{6}{7}$ of £420   **b** $\frac{3}{8}$ of 320 counters   **c** $\frac{2}{5}$ of 365 days

d $\frac{5}{12}$ of 60 minutes   **e** $\frac{2}{9}$ of £63   **f** $\frac{6}{11}$ of 44 litres

4 In the London Marathon, 32 000 runners took part, of which $\frac{3}{16}$ were female. How many women ran the race?

(FS) 5 A house is valued at £130 000. A couple take out a mortgage for $\frac{3}{5}$ of this amount. How much is the mortgage?

(FS) 6 A lottery syndicate has 25 shares. Jane has three of them. How much will Jane receive out of a win of £15 700?

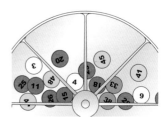

7 Write each pair of fractions with a common denominator. Then put the correct sign, > or <, between them.

a $\frac{4}{5}, \frac{7}{9}$ **b** $\frac{2}{7}, \frac{3}{8}$ **c** $\frac{1}{4}, \frac{2}{9}$ **d** $\frac{5}{8}, \frac{3}{5}$

8 Which of the amounts in each pair is larger?

a $\frac{6}{7}$ of 84 or $\frac{5}{6}$ of 90   **b** $\frac{3}{8}$ of 44 or $\frac{2}{9}$ of 72

c $\frac{3}{10}$ of 85 or $\frac{3}{11}$ of 88   **d** $\frac{5}{12}$ of 96 or $\frac{5}{7}$ of 63

9 Which of the fractions below is closest to $\frac{1}{4}$?

a $\frac{3}{11}$ **b** $\frac{3}{8}$ **c** $\frac{5}{18}$ **d** $\frac{3}{10}$

10 210 000 candidates entered a GCSE examination. Of these, $\frac{3}{8}$ were tested by Exam Board A, $\frac{4}{7}$ by Exam Board B and the rest by Exam Board C.

a How many candidates took Exam Board A's paper?

b How many candidates took Exam Board B's paper?

c How many candidates took Exam Board C's paper?

Revision of adding and subtracting fractions

Work these out.

a $\frac{3}{7} + \frac{2}{9}$ **b** $\frac{7}{10} - \frac{3}{8}$ **c** $2\frac{5}{6} + 1\frac{7}{15}$

To add fractions with different denominators, you need to find the lowest common denominator. This is the same as the lowest common multiple (LCM) of the denominators. Always simplify answers to their lowest terms and convert to mixed numbers if necessary.

a $\frac{3}{7} + \frac{2}{9} = \frac{27}{63} + \frac{14}{63} = \frac{41}{63}$ The LCM of 7 and 9 is 63.

b $\frac{7}{10} - \frac{3}{8} = \frac{28}{40} - \frac{15}{40} = \frac{13}{40}$ The LCM of 10 and 8 is 40.

c $2\frac{5}{6} + 1\frac{7}{15} = 2 + \frac{5}{6} + 1 + \frac{7}{15}$

$= 3 + \frac{5}{6} + \frac{7}{15}$ Work out the whole-numbers separately. $2 + 1 = 3$

$\frac{5}{6} + \frac{7}{15} = \frac{25}{30} + \frac{14}{30} = \frac{39}{30} = 1\frac{9}{30}$ The LCM of 6 and 15 is 30.

$3 + 1\frac{9}{30} = 4\frac{9}{30}$ Add the whole-number part to the fraction part.

$4\frac{9}{30} = 4\frac{3}{10}$ Simplify to lowest terms.

Exercise 14G

1 There are 1500 pupils in a school.
$\frac{2}{5}$ of them come by bus, $\frac{1}{3}$ walk and the rest come by car.

a What fraction come by car?

b How many pupils come by bus?

2 In a small village there are 620 children, of which $\frac{9}{20}$ are boys. $\frac{5}{9}$ of the boys go to high school.

a How many boys go to high school?

b How many girls are there in the village?

3 On checking his emails Mr Smith finds that $\frac{1}{3}$ of them are junk mail, $\frac{3}{8}$ are about business and the rest are personal. What fraction of his emails are personal?

4 Mr Berry collects cheese-boxes. He has 630.
Of these, $\frac{4}{9}$ are square and $\frac{1}{18}$ are rectangular.

a What fraction of the collection are not square or rectangular?

b How many square boxes are there in his collection?

5 Mr Giles has a large farm. He uses $\frac{3}{8}$ of the fields for crops, $\frac{1}{6}$ to graze cattle and he leaves the rest lying fallow. Of the fields used for crops $\frac{2}{3}$ are planted with wheat.

 a What fraction of the farm is lying fallow?

 b There are 240 acres on the farm. How many acres are planted with wheat?

6 Two packets of nails weigh $3\frac{3}{5}$ kg and $2\frac{1}{6}$ kg.

 How much do they weigh altogether?

7 **a** Work these out.

 i $\frac{1}{2}+\frac{1}{4}$ **ii** $\frac{1}{2}+\frac{1}{4}+\frac{1}{8}$ **iii** $\frac{1}{2}+\frac{1}{4}+\frac{1}{8}+\frac{1}{16}$

 b What is the answer to the infinite sum below?

 $$\frac{1}{2}+\frac{1}{4}+\frac{1}{8}+\frac{1}{16}+\frac{1}{32}+\frac{1}{64}+\frac{1}{128}+\frac{1}{256}+\frac{1}{512}+\ldots$$

8 Work these out. Simplify your answers to lowest terms and change any improper fractions into mixed numbers.

 a $\frac{1}{12}+\frac{2}{3}$ **b** $\frac{3}{4}+\frac{5}{8}$ **c** $\frac{2}{7}-\frac{1}{8}$ **d** $\frac{11}{15}-\frac{1}{6}$

 e $\frac{7}{9}-\frac{3}{8}$ **f** $\frac{17}{20}+\frac{7}{15}$ **g** $\frac{9}{10}-\frac{2}{3}$ **h** $\frac{3}{7}+\frac{5}{6}$

9 Work these out. Simplify your answers to lowest terms and change any improper fractions into mixed numbers.

 a $2\frac{7}{15}+1\frac{2}{3}$ **b** $2\frac{5}{12}-1\frac{5}{8}$ **c** $3\frac{11}{20}+1\frac{5}{6}$ **d** $3\frac{9}{10}+1\frac{1}{6}$

 e $4\frac{4}{7}+1\frac{7}{8}$ **f** $3\frac{6}{7}+1\frac{1}{2}$ **g** $4\frac{1}{3}-2\frac{1}{6}$ **h** $7\frac{8}{15}+3\frac{5}{12}$

10 A tin contains $3\frac{3}{8}$ litres of syrup.

 Jamie pours out $1\frac{3}{4}$ litres of syrup.

 How many litres are left?

11 Find the perimeter of each shape.

 a

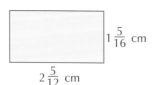

$2\frac{5}{12}$ cm, $1\frac{5}{16}$ cm

 b

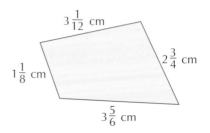

$3\frac{1}{12}$ cm, $2\frac{3}{4}$ cm, $1\frac{1}{8}$ cm, $3\frac{5}{6}$ cm

Revision of algebra

Example 6

Expand and then simplify each expression.

a $2(x - 2) + 3(x + 1)$ **b** $2(3x - 1) - 3(x - 2)$

Expand the brackets and then collect like terms.

a $2(x - 2) + 3(x + 1) = 2x - 4 + 3x + 3 = 2x + 3x - 4 + 3 = 5x - 1$

b $2(3x - 1) - 3(x - 2) = 6x - 2 - 3x + 6 = 6x - 3x - 2 + 6 = 3x + 4$

Example 7

Expand and then simplify each expression.

a $x(x - y) + 2x(y + 3x)$ **b** $x(x + 2y) - y(2x - y)$

a Expand each bracket and then collect like terms together:

$x(x - y) + 2x(y + 3x) = x^2 - xy + 2xy + 6x^2 = x^2 + 6x^2 - xy + 2xy = 7x^2 + xy$

b $x(x + 2y) - y(2x - y) = x^2 + 2xy - 2yx + y^2 = x^2 + y^2$

Note that $2xy$ and $2yx$ are the same expression.

Example 8

Solve these equations.

a $3(x - 1) + 2(2x - 1) = 23$ **b** $3(x + 1) = 2(x - 2) + 12$

a $3(x - 1) + 2(2x - 1) = 23$

$3x - 3 + 4x - 2 = 23$ Expand the brackets.

$7x - 5 = 23$ Collect terms.

$7x = 28$ Add 5 to both sides.

$x = 4$ Divide by 7.

b $3(x + 1) = 2(x - 2) + 12$

$3x + 3 = 2x - 4 + 12$ Expand the brackets.

$3x - 2x = -4 + 12 - 3$ Collect letter terms on the left and numbers on the right of the equals sign.

$x = 5$ Add or subtract the like terms.

Exercise 14I

1 Expand and simplify each expression.

a $2(x - 4) + 3(x + 2)$ **b** $3(x - 1) - 2(x + 3)$

c $4(2x + 5) - 3(3x - 2)$ **d** $4(x + 5) - 2(x - 10)$

e $3(3x + 1) + 2(2x - 7)$ **f** $3(x - 5) - 3(2x - 1)$

g $x(x + 2y) - x(3y - x)$ **h** $x(2x - y) - 2x(x + 2y)$

i $3x(2x - 3y) - x(3x + 2y)$ **j** $3x(2x - 3y) + x(y + 5x)$

2 Solve each equation.

a $2(x - 4) + 3(x + 2) = 18$ **b** $3(x - 1) - 2(x + 3) = 19$

c $4(x + 1) - 2(x - 3) = 20$ **d** $2(x + 3) + 4(x - 3) = 15$

e $3(x + 2) - 4(x - 3) = 19$ **f** $5(x + 2) - 6(x + 1) = 6$

g $2(3x - 1) + 2(x + 3) = 4$ **h** $5(2x + 1) + 4(x - 3) = 21$

i $4(2 - 3x) + 2(2x - 1) = 14$ **j** $2(4x + 2) + 3(2x + 7) = 4$

3 Solve each equation.

a $5(x - 1) = x + 3$ **b** $9(x - 3) = 3(x + 1)$

c $4(x + 2) + 5 = 2(x - 1) + 18$ **d** $6(x - 2) + 9 = 6(2x - 2)$

e $3(x - 1) = 4(x + 2) - 9$ **f** $3(2x - 1) = 2(4x - 3) - 2$

4 The angles of a triangle, in degrees, are $3x$, $2x + 15$ and $4x - 60$.
Calculate the value of x.

5 The perimeter of this rectangle is 22 cm.

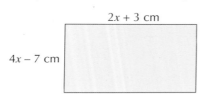

2x + 3 cm

4x − 7 cm

Calculate the value of x.

GCSE-type questions

1 **a** Work these out.   **i** $5 - 12$ **ii** -4×-7

     b Work out $\frac{5}{8} - \frac{2}{5}$.

2 Gary's mark in a mathematics test was 42 out of 50.
What was Gary's mark as a percentage?

3 **a** Work out $28 - (15 + 7)$.

     b Insert brackets to show the order in which you would do this calculation.
     $16 - 4 \times 2 = 8$

4 Explain how you would find the smallest and the largest of these three numbers.

$$0.21 \qquad \frac{5}{19} \qquad 22\%$$

5 Ken buys 1.5 kilograms of potatoes at 84 pence per kilogram.

He also buys some carrots at 76 pence per kilogram.

His total bill is £2.21.

What is the mass of the carrots that he buys?

(FS) **6** **a** What is 42% of £258.

b What percentage is £45 of £180?

7 Mrs Ball has double glazing installed in her house. It cost £9000.

To pay for the double glazing, she must first pay a deposit of 12% of the total price.

The rest of the cost is paid in 10 equal instalments

How much is one of these equal payments?

8 You know that $64 \times 135 = 8640$.

Explain how you would use this fact to work out:

a 64×270 **b** $864 \div 0.64$.

9 Work out the value of $\frac{3}{5} \times \frac{7}{12}$.

Give your answer as a fraction in its simplest form.

(FS) **10** Tebor sells DVDs.

He sells each DVD for £11.80 plus VAT at 20%.

If he sells 540 DVDs, how much money will he receive?

11 **a** Expand $x(x + 3)$.

b Expand and simplify $2(3x + 4) + 3(4x - 3)$.

12 Expand and simplify $4(2x - 3) + 3(2x + 4)$.

13 Find the lowest common multiple of 24 and 36.

14 The sizes of the angles in this triangle, in degrees, are $2x$, $4x$ and $2x + 4$.

Work out the size of each angle.

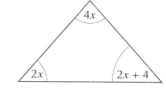

15 ABC is a triangle.

Giving reasons, state the size of the angle labelled:

a y **b** x **c** t.

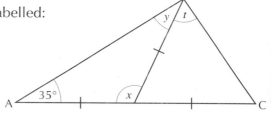

16 Draw the graph of $y = x^2 - 4x - 2$. Draw your coordinate axes and choose suitable ranges of values.

17 The mass of a box was 35 kg, correct to the nearest kg.

    a Write down the smallest possible mass of the box.

    b Write down the largest possible mass of the box.

 18 The nth term of a sequence is $n^2 + 2$.

    Trevor says: 'The nth term of this sequence is always a prime number when n is odd.'

    Is Trevor correct? Explain your answer.

19 The table gives information about the food sold one day in a small café.

Food	Frequency	Size of angle
Soup and crusty bread	10	40°
Fish and chips	35	
Pizza	20	
Pie and chips	25	

    a Copy and complete the table.

    b Draw a pie chart to show this information.

 20 Boyd recorded the times, in minutes, it took him to mend laptops.

    The table shows his results.

Time, t (minutes)	Frequency
$0 < t \leqslant 10$	8
$10 < t \leqslant 20$	28
$20 < t \leqslant 30$	23
$30 < t \leqslant 40$	16
$40 < t \leqslant 50$	14
$50 < t \leqslant 60$	11

    Calculate an estimate for the mean time taken to repair a laptop.

21 Emily has a bag of sweets.

    12 of the sweets are jellies.

    18 of the sweets are toffees.

    a Write down the ratio of the number of jellies to the number of toffees.

      Give this ratio in its simplest form.

    b Jack has some mints and chews in a box.

      The total number of sweets in the box is 63.

      The ratio of the number of mints to the number of chews is 2 : 5.

      How many of the sweets in Jack's box are: **i** mints   **ii** chews?

 22 The diameter of this semicircular card is 10 cm.

    Work out the perimeter of the card. Give your answer correct to one decimal place.

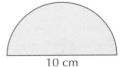

10 cm

Glossary

adjacent   Next to; in trigonometry, the side between the right angle and the specified angle.

appropriate degree of accuracy   A degree of accuracy that is suitable for the way the number is being used.

cancel   Simplify, by dividing numerator and denominator of a fraction by a common factor.

composite shape   A shape composed of two or more basic shapes.

compound unit   A unit that is made up of two other units, such as kilometres per hour (km/h) or grams per cubic centimetre (g/cm^3).

conversion   Change; changing a unit into a smaller or larger one.

conversion factor   The number by which a unit is multiplied or divided, to change it to a different unit.

cosine   The ratio of the length of the side that is next to an acute angle, to the length of the hypotenuse, in a right-angled triangle.

cross-section   A cut across a 3D shape. The face that is exposed when a 3D shape is cut. For a prism, a cut across the shape, perpendicular to its length.

curved surface area   The area of a curved surface, such as the curved part of a cylinder or the whole surface of a sphere.

decrease   Reduction; make smaller.

density   The mass of a substance divided by its volume.

discontinuous graph   A graph in which the values of one of the variables has fixed values for different values of the other variable; the graph does not form a continuous range of values, they cannot be joined up.

distance–time graph   A graph that displays a journey, based on the distance travelled and the time taken.

expand   Multiply out (terms with brackets).

exponential growth graph   A graph that shows the relationship between two variables, one of which is expressed as a power.

exterior angle   The angle formed outside a 2D shape, when a side is extended beyond the vertex.

factorise   Take out a common factor and write an expression as a bracketed term with a multiplying factor in front.

hectare   A measure of area, equal to 10 000 square metres.

hypotenuse   The longest side in a right-angled triangle, always opposite the right angle.

increase   Make larger.

interior angle   The inside angle between two adjacent sides of a 2D shape, at a vertex.

inverse trigonometric function   An expression that can be used to find the size of an angle, given the value of its trigonometric ratio; $\sin^{-1}$, $\cos^{-1}$ and $\tan^{-1}$.

invert   Exchange the numerator and denominator of a fraction.

irregular polygon   A polygon in which the sides are not all equal, and the angles are not all equal.

lender   Someone who allows someone else (the borrower) to borrow money from them.

linear equation   An equation in which the highest power of any variable is 1; an equation that can be represented by a straight line on a graph.

multiplier   A number that is used to find the result of increasing or decreasing an amount by a percentage.

negative correlation   A relationship between two sets of data, in which the value of one variable increases as the value of the other variable decreases.

negative power   The reciprocal of a positive power; for example, $10^{-1} = \frac{1}{10}$, $10^{-2} = \frac{1}{10^2} = \frac{1}{100}$.

no correlation   A lack of any sort of relationship between two sets of data.

open cylinder   A cylinder that has no ends.

opposite   The side that is opposite to a specified acute angle, in a right-angled triangle.

original value   The value before a change (increase or decrease).

polygon   A closed 2D shape with straight sides.

positive correlation   A relationship between two sets of data, in which the value of one variable increases as the value of the other variable increases.

prism   A 3D shape that has the same shape of cross-section wherever it is cut perpendicular to its length.

Pythagoras   A Greek philosopher who studied mathematics.

Pythagoras' theorem   The rule that, in any right-angled triangle, the square of the hypotenuse is equal to the sum of the squares on the other two sides.

Pythagorean triple   Three whole numbers that satisfy the relation $a^2 + b^2 = c^2$.

quadratic   Having terms involving one or two variables, and constants, such as $x^2 - 3$ or $y^2 + 2y + 4$ where the highest power of the variable is two.

rate   A comparison of how one variable changes in relation to another.

rearrange   Put into a different order, to simplify.

regular polygon   A polygon in which the sides are all equal, and the angles are all equal.

scatter graph   A diagram showing corresponding values between two sets of data.

semi-tessellation   A tessellation made up of two or more repeating 2D shapes.

shorter side   In a right-angled triangle, either of the sides that is not the hypotenuse.

similar   Two shapes are similar if one is an enlargement of the other; angles in the same position in both shapes are equal to each other.

simple interest   Money that a borrower pays a lender, for allowing them to borrow money.

simultaneous equations   Two equations that are both true for the same set of values for their variables.

sine   The ratio of the length of the side that is opposite an acute angle, to the length of the hypotenuse, in a right-angled triangle.

speed   The rate at which an object is travelling.

standard form   A way of writing a number as $a \times 10^n$, where $1 \leqslant a < 10$ and n is a positive or negative integer.

step graph   A graph in which the values of one of the variables has fixed values for different values of the other variable; the graph does not form a continuous range of values, they cannot be joined up.

subject   The variable on the left-hand side of the equals (=) sign in a formula or equation.

surface area   The area of a surface, usually of a 3D shape.

tangent   The ratio of the length of the side that is opposite an acute angle, to the length of the side that is next to the same angle, in a right-angled triangle.

tessellate   Fit together, a repeating 2D shape, with no gaps.

tessellation   A pattern formed by a repeating 2D shape.

time graph   A graph that shows the values of a variable, based on the progression of time.

time-series graph   A graph that is based on a time sequence.

total surface area   The total area of all of the surfaces of a 3D shape.

two-way table   A table that records values that depend on two sets of criteria.

unit price   The cost of one unit, such as a kilogram, litre or metre, of something.

variable   A letter that stands for a quantity that can take various values.

Index

William Collins's dream of knowledge for all began with the publication of his first book in 1819. A self-educated mill worker, he not only enriched millions of lives, but also founded a flourishing publishing house. Today, staying true to this spirit, Collins books are packed with inspiration, innovation and practical expertise. They place you at the centre of a world of possibility and give you exactly what you need to explore it.

Collins. Freedom to teach.

Published by Collins
An imprint of HarperCollins*Publishers*
The News Building
1 London Bridge Street
London
SE1 9GF

Browse the complete Collins catalogue at
www.collins.co.uk

© HarperCollins*Publishers* Limited 2014

10 9 8 7

ISBN-13 978-0-00-753778-5

The authors Kevin Evans, Keith Gordon, Chris Pearce, Trevor Senior and Brian Speed assert their moral rights to be identified as the authors of this work.

British Library Cataloguing in Publication Data
A catalogue record for this publication is available from the British Library.

Commissioned by Katie Sergeant
Project managed by Elektra Media Ltd
Development edited by Lindsey Charles
Developed and copy-edited by Joan Miller
Edited by Helen Marsden
Proofread by Amanda Dickson
Illustrations by Ann Paganuzzi, Nigel Jordan and Tony Wilkins
Typeset by Jouve India Private Limited
Cover design by Angela English

Printed and bound by Grafica Veneta S.p.A.

Acknowledgements
The publishers wish to thank the following for permission to reproduce photographs. Every effort has been made to trace copyright holders and to obtain their permission for the use of copyright materials. The publishers will gladly receive any information enabling them to rectify any error or omission at the first opportunity.

(t = top, c = centre, b = bottom, r = right, l = left)

Cover Hupeng/Dreamstime.com, p 6 godrick/Shutterstock, p 8 mangostock/Shutterstock, p 14 WDG Photo/Shutterstock, p 15 vovan/Shutterstock, p 16 Andresr/iStock, p 18t David Steele/Shutterstock, p 18b Laura Stone/Shutterstock, p 21 Mircea BEZERGHEANU/Shutterstock, p 22–23 isak55/Shutterstock, p 24 SSPL/Getty Images, p 36 Correcaminos112/Shutterstock, p 40–41 Denys Prykhodov/Shutterstock, p 42 Horia Bogdan/Shutterstock, p 58–59 Scott E. Feuer/Shutterstock, p 60 Iakov Kalinin/Shutterstock, p 63 Martin Good/Shutterstock, p 64l Anastasios71/Shutterstock, p 64r Tatiana Popova/Shutterstock, p 69 Nemar74/Shutterstock, p 74l DDCoral/Shutterstock, p 74r Posmetukhov Andrey/Shutterstock, p 82–83 guentermanaus/Shutterstock, p 84 Natursports/Shutterstock, p 98–99 Rido/Shutterstock, p 100 mmmm/Shutterstock, p 114–115 MHA-Archive/Alamy, p 166 Nils Z/Shutterstock, p 128–129 GrandeDuc/Shutterstock, p 130 Nicholas Rjabow/Shutterstock, p 140–141 Mikhail Bakunovich/Shutterstock, p 142 leungchopan/Shutterstock, p 145 B. Franklin/Shutterstock, p 146 Just Keep Drawing/Shutterstock, p 148 Pan Xunbin/Shutterstock, p 149 stevenku/Shutterstock, p 151 Tracey Helmboldt/Shutterstock, p 154 Silberkorn/Shutterstock, p 156 Dream79/Shutterstock, p 159t matka_Wariatka/Shutterstock, p 159b Rick P Lewis/Shutterstock, p 160–161 severija/Shutterstock, p 162 Aleksandra Pikalova/Shutterstock, p 180–181 ValeStock/Shutterstock, p 181tl Evgeny Karandaev/Shutterstock, p 181tcl Louise Cukrov/Shutterstock, 181tcr Jon Le-Bon/Shutterstock, p 181tr donydony/Shutterstock, p 181bl dcwcreations/Shutterstock, p 181bcl Louise Cukrov/Shutterstock, p 181bcr Jon Le-Bon/Shutterstock, p 181br K. Miri Photography/Shutterstock, p 182 MrOK/Shutterstock, p 194–195 SasinT/Shutterstock, p 196 stockyimages/Shutterstock, p 198 Ivan Kruk/Shutterstock, p 199 Igor Sh/Shutterstock, p 202 Pixelbliss/Shutterstock, p 203 Max Sudakov/Shutterstock, p 205 Diana Taliun/Shutterstock, p 206 s-ts/Shutterstock, p 207 Diana Taliun/Shutterstock, p 208 Stu Porter/Shutterstock, p 210–211 Canadapanda/Shutterstock, p 212 Jeffrey M. Frank/Shutterstock, p 228–229 chrisdorney/Shutterstock, p 229 Lebrecht Music and Arts Photo Library/Alamy, p 230 Sabphoto/Shutterstock.